BRIDGE ENGINEERING

Edited by **Hamid Yaghoubi**

Bridge Engineering
http://dx.doi.org/10.5772/intechopen.70024
Edited by Hamid Yaghoubi

Contributors

Selcuk Bas, Javier Fernando Jiménez-Alonso, Andres Saez, Ján Benčat, Robert Kohar, Daniel Vaz, Raquel Albuquerque Almeida, A.R. Janeiro Borges, Roszilah Hamid, Khairullah Yusof, Faham Tahmasebinia, Samad Mohammad Ebrahimzadeh Sepasgozar, Hamid Yaghoubi

First published in London, United Kingdom, 2018 by IntechOpen
IntechOpen is the global imprint of INTECHOPEN LIMITED, registered in England and Wales, registration number: 11086078, The Shard, 25th floor, 32 London Bridge Street
London, SE19SG – United Kingdom
Printed in Croatia

British Library Cataloguing-in-Publication Data
A catalogue record for this book is available from the British Library

Additional hard copies can be obtained from orders@intechopen.com

Bridge Engineering, Edited by Hamid Yaghoubi
p. cm.
Print ISBN 978-1-78923-104-5
Online ISBN 978-1-78923-105-2

For EU product safety concerns:
IN TECH d.o.o., Prolaz Marije Krucifikse Kozulić 3, 51000 Rijeka, Croatia,
info@intechopen.com or visit our website at intechopen.com.

Meet the editor

Dr. Hamid Yaghoubi is the director of Iran Maglev Technology (IMT). He became the Iran top researcher in 2010. In this regard, he was awarded by the Iranian president; the Iranian Minister of Science, Research and Technology; and the Iranian Minister of Information and Communication Technology. He became the 2011 and 2012 Outstanding Reviewer for the *Journal of Transportation Engineering* (JTE), *American Society of Civil Engineers* (ASCE), USA. One of his journal papers became the 2011 Top Download Paper for JTE. He received the ICCTP2011 Award for the 11th International Conference of Chinese Transportation Professionals (ICCTP2011), ASCE. He is an assistant chief editor and an editorial board member for some journals. He has been a reviewer for the majority of journals, books and conferences. He has also been an editor for some books. He has cooperated with hundreds of international conferences as a chairman, a keynote speaker, a chair of session, a publication chair, and a member of committees, including scientific, organizing, steering, advisory, technical program, and so on. He is also a member of several international committees.

Contents

Preface

A bridge is a structure built to span the physical obstacles without closing the way underneath, such as a body of water, valley, or road, for the purpose of providing passage over the obstacle. Bridge engineering is an engineering discipline branching from civil engineering that involves the planning, design, construction, operation, and maintenance of bridges to ensure safe and effective transportation of vehicles, people and goods. The book *Bridge Engineering* includes the main topics and the basic principles of bridge engineering and provides the full scope of current information necessary for effective and cost-conscious contemporary bridge. It reflects new engineering and building developments, the most current design methods, and the latest industry standards and policies. It provides a comprehensive overview of the significant characteristics for bridge engineering. It highlights the recent advancements, requirements and improvements and details of the latest techniques in the global market. It contains a collection of the latest research developments on the bridge engineering. It comprehensively covers the basic theory and practice in sufficient depth to provide a solid grounding to bridge engineers. It helps readers to maximize effectiveness in all facets of bridge engineering. This professional book as a credible source and a valuable reference can be very applicable and useful for all professors, researchers, engineers, practicing professionals, trainee practitioners, students, and others who are interested in the bridge projects. The book *Bridge Engineering* consists of seven chapters.

Chapter 1 reviews the maglev-elevated guideways. Contrary to the traditional railway trains, there is no direct contact between the maglev vehicle and its guideway. These vehicles travel along the magnetic fields that are established between the vehicle and its guideway. In contrary to the traditional railroad tracks, there is no need for ballast, sleeper, rail pad, and rail fastenings to stabilize the rail gauge. Basically, there are two main elements in a maglev system including its vehicle and the guideway. The guideway is the structure that maglev vehicles move over it and are supported and guided by it. It is the main element in maglev system and holds a big share of costs for the system. The guideway consists of superstructures and substructures. In fact, a guideway consists of a beam and two levitation rails. Guideways can be constructed at grade or elevated including columns with concrete, steel, or hybrid beams. Concrete guideway girders can be reinforced or prestressed. Majority of the existing maglev guideways are elevated and completely built on bridge. Guideway provides guidance on the movement of the vehicle to support the vehicle load and transfer the load to the ground. The loading of the maglev vehicle is an important parameter in the practical application. Guideway girder is evaluated for different load cases. During the past three decades, different guideways have been developed, constructed and tested.

Chapter 2 explores the possibility of designing and constructing a super-long-span bridge with new materials in 2050. The features of this structure include the combination of suspension and cable-stayed bridges, application of carbon fibre materials, extension of deck width, and pretension techniques. Linear static analysis, dynamic analysis and theoretical analysis are conducted under different loading cases. In linear static analysis, the stresses under critical load combinations are smaller than the ultimate strength of the materials. However, the maximum deflection under the dead and wind load combination exceeds the specified serviceability limit.

Chapter 3 utilizes the artificial neural network (ANN) and multiple regression analysis (MRA) to model bridge condition rating based on the limited number of data sets. Bridge condition rating models are developed based on very limited inspection data records using MRA and ANN techniques. Since the data sets are limited, utilizing all the available data sets by handling missing data with the appropriate methods is expected to improve the performance of the models. Five methods are then used to handle the missing bridge component condition rating data. Three commonly used methods and two new methods are explored in this study. It seems that the performance of the model using data sets after handling the missing bridge component data to fill the gaps in the range scales of the bridge condition rating improved the performance of the model. Furthermore, a handling method that substitutes the missing data of bridge component ratings with available bridge rating data is the most reasonable.

In *Chapter 4*, structural identification (St-Id) is detailed and implemented to estimate the performance of the Bosphorus Suspension Bridge. In addition, certain efforts from finite element modeling (FEM) to utilization and decision-making for performance prediction are given based on each step of St-Id. In general, the St-Id concept is divided into two main parts: experimental and numerical investigations. Due to the high cost and time limitation for testing of long-span bridges, the most effective solution to the experimental research is SHM system (SHMs). For this purpose, the SHMs of the Bosphorus Bridge is considered and is more detailed on the basis of the bridge. As to numerical investigation, finite element modeling provides an extended solution from analysis to model updating of the bridges. Taking the considerations of the St-Id concept and the outcomes from the Bosphorus Bridge into account, structural performance of the bridge under extreme wind load and multipoint earthquake motion is estimated.

Chapter 5 evaluates the recent advances in the serviceability assessment of footbridges under pedestrian-induced vibrations. Current international guidelines determine the effect of pedestrians on footbridges through an equivalent harmonic load. However, the dynamic response of footbridges obtained according to these standards differs from the values recorded experimentally. In order to overcome this issue, a new modeling framework has been recently proposed by several researchers. This novel approach allows considering more accurately three key aspects: (i) the inter- and intra-subject variability, (ii) the pedestrian-structure interaction and (iii) the crowd dynamics. For this purpose, different crowd-structure interaction models have been developed. Despite the large numbers of proposals, all of them share the same scheme: the crowd-structure interaction is simulated by linking two sub-models, namely, (i) a pedestrian-structure interaction sub-model and (ii) a crowd sub-model. Furthermore, the variability of the pedestrian's behavior may be taken into account through the assumption that the model parameters are random variables. In this chapter, a summary of the state of the art of this new modeling framework is presented with a special emphasis in a case study, where the crowd-structure interaction model is used to simulate the lateral lock-in phenomenon on a real footbridge.

In *Chapter 6*, several aerodynamic phenomena of relevance to long-span bridges are classified and discussed. For certain cases, codes of practice recommend wind tunnel tests. The reader is introduced to these as well as to numerical simulations, which are currently gaining increasing importance. Next, measures for attenuating susceptibility for undesirable dynamic responses are reviewed. The chapter ends with a discussion of the Vila Real Bridge deck-section based on wind tunnel tests and numerical simulations. The aerodynamics was effectively improved with geometrically subtle modifications that were proposed and adopted still in the design phase.

Chapter 7 reviews the analysis of problems of highway and rail bridge dynamic response to moving traffic loads. Bridge vibration analyses comprise solution of many interdisciplinary problems. During the two last centuries, these problems have been studied by theoretical, numerical, and experimental way by many investigators. Therefore, the present chapter contains only the basic approaches for solving the complex problem of bridges subjected to dynamic loading.

Dr. Hamid Yaghoubi
Director of Iran Maglev Technology (IMT)
Iran

Chapter 1

Introductory Chapter: Modern Bridges

Hamid Yaghoubi

Additional information is available at the end of the chapter

http://dx.doi.org/10.5772/intechopen.74722

1. Introduction

Bridge engineering is an engineering discipline branching from civil engineering that involves the planning, design, construction, operation, and maintenance of bridges to ensure safe and effective transportation of vehicles, people, and goods. Among transportation of vehicles, maglev (magnetic levitation) systems have become a focus of the worldwide transportation industries. The need for rapid transit systems has become vital in both urban and intercity travels. Application of magnetically levitated trains has attracted numerous transportation industries throughout the world. Contrary to the traditional railway trains, there is no direct contact between the maglev vehicle and its guideway. These vehicles travel along the magnetic fields that are established between the vehicle and its guideway. There are already many countries attracted to maglev systems (see **Figure 1**) [1–16].

2. Maglev elevated guideways

In maglev guideways, contrary to the traditional railroad tracks, there is no need to ballast, sleeper, rail pad and rail fastenings to stabilize the rail gauge. Basically, there are two main elements in a maglev system, including its vehicle and the guideway. The guideway is the structure that maglev vehicles move over it and are supported and guided by it. It is the main element in maglev system and holds big share of costs for the system. It is vital for maglev trains. Guideway consists of superstructures and substructures. In fact, a guideway consists of a beam (girder) and two levitation (guidance) rails. Guideways can be constructed at grade (ground-level) or elevated including columns with concrete, steel, or hybrid beams. Concrete guideway girders can be as reinforced or prestressed. Majority of the existing maglev guideways are elevated and completely built on bridge (see **Figure 2**). Guideway provides

Figure 1. China maglev guideway.

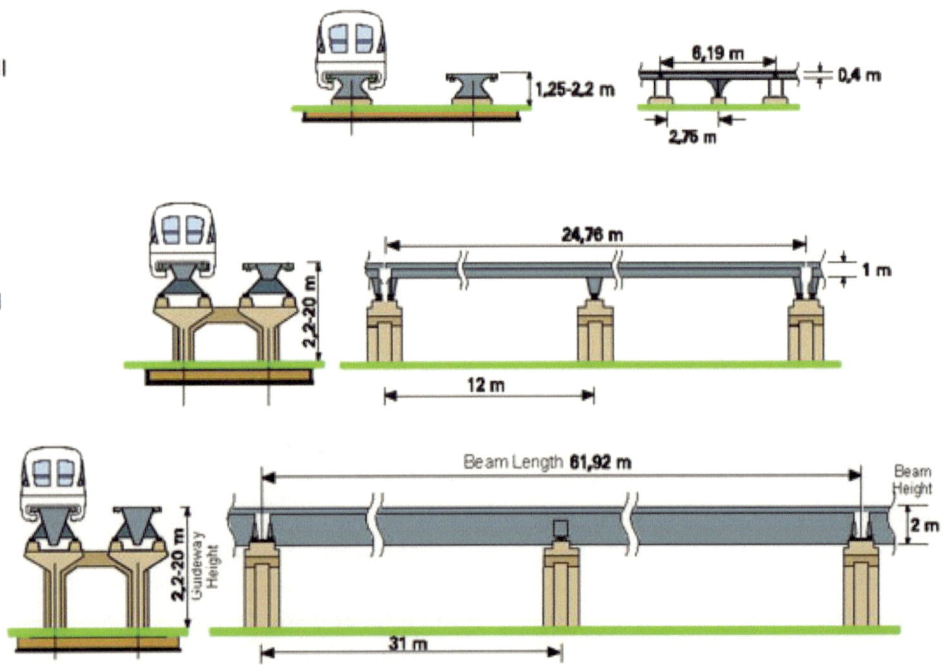

Figure 2. Standard guideway types.

guidance for the movement of the vehicle, to support the vehicle load, and to transfer the load to the ground. The loading of the maglev vehicle is an important parameter in the practical application. It is related to the magnetic forces. Guideway girder is evaluated for different

http://dx.doi.org/10.5772/intechopen.74722

load cases. Magnetic forces are generated by the maglev vehicle and cause structural loading that transmits to the guideway. This can happen while such a vehicle is stationary or in motion. There are a variety of designs for maglev systems, and engineers keep revealing new ideas about such systems. Many systems have been proposed in different parts of the world, and a number of corridors have been selected and researched. During the past three decades, different guideways have been developed, constructed, and tested [1–16].

Acknowledgements

This work was performed by Iran Maglev Technology (IMT).

Author details

Hamid Yaghoubi

Address all correspondence to: info@maglev.ir

Iran Maglev Technology (IMT), Tehran, Iran

References

[1] Iran Maglev Technology (IMT). 2008. www.maglev.ir, Tehran, Iran

[2] Yaghoubi H. Magnetically Levitated Trains, Maglev. Tehran, Iran: Pooyan Farnegar Publisher; 2008. ISBN: 978-600-5085-05-1

[3] Yaghoubi H, Sadat Hoseini M. Mechanical assessment of maglev vehicle—A proposal for implementing maglev trains in Iran. The ASME 10th Biennial Conference on Engineering Systems Design and Analysis (ESDA); Yeditepe University, Istanbul, Turkey, Vol. 2. 2010. pp. 299-306. ISBN: 978-0-7918-4916-3

[4] Yaghoubi H, Ziari H. Assessment of structural analysis and design principles for maglev guideway: A case-study for implementing low-speed maglev systems in Iran. The 1st International Conference on Railway Engineering, High-speed Railway, Heavy Haul Railway and Urban Rail Transit; Beijing Jiaotong University; Beijing, China: China Railway Publishing House; 2010. pp. 15-23. ISBN: 978-7-113-11751-1

[5] Behbahani H, Yaghoubi H. Procedures for safety and risk assessment of maglev systems: A case-study for long-distance and high-speed maglev project in Mashhad-Tehran route. The 1st International Conference on Railway Engineering, High-speed Railway, Heavy Haul Railway and Urban Rail Transit; Beijing Jiaotong University; Beijing, China: China Railway Publishing House; 2010. pp. 73-83. ISBN: 978-7-113-11751-1

[6] Yaghoubi H. The most important advantages of magnetically levitated trains. Towards Sustainable Transportation Systems. Proceedings of the 11th International Conference

of Chinese Transportation Professionals (ICCTP2011); 2011; Nanjing, China, American Society of Civil Engineers (ASCE) Publisher. p. 3974-3986. ISBN: 978-0-7844-1186-5

[7] Yaghoubi H, Ziari H. Development of a maglev vehicle/guideway system interaction model and comparison of the guideway structural analysis with railway bridge structures. ASCE Journal of Transportation Engineering. 2011;**137**(2):140-154

[8] Yaghoubi H, Barazi N, Kahkeshan K, Zare A, Ghazanfari H. Technical comparison of maglev and rail rapid transit systems. The 21st International Conference on Magnetically Levitated Systems and Linear Drives (MAGLEV 2011); 2011; Daejeon Convention Center, Daejoen, Korea

[9] Yaghoubi H, Rezvani MA. Development of maglev guideway loading model. ASCE Journal of Transportation Engineering. 2011;**137**(3):201-213

[10] Behbahani H, Yaghoubi H, Rezvani MA. Development of technical and economical models for widespread application of magnetic levitation system in public transport. International Journal of Civil Engineering (IJCE). 2012;**10**(1):13-24

[11] Yaghoubi H, Barazi N, Aoliaei MR. Maglev. Chapter 6. Infrastructure Design, Signalling and Security in Railway; University Campus STeP Ri, Rijeka, Croatia: InTech; 2012. pp. 123-176. ISBN: 978-953-51-0448-3

[12] Yaghoubi H. Practical applications of magnetic levitation technology. Final Report. Tehran, Iran: Iran Maglev Technology (IMT); 2012. Available from: http://www.maglev.ir/eng/documents/reports/IMT_R_22.pdf

[13] Yaghoubi H, Keymanesh MR. Design and evaluation criteria for stations of magnetically levitated trains. Journal of Civil Engineering and Science (JCES). 2013;**2**(2):72-84. ISSN: 2227-4634 (print), ISSN: 2227-4626 (online)

[14] Yaghoubi H. The most important maglev applications. Journal of Engineering. 2013;**2013**:19 pages. DOI: 10.1155/2013/537986. ISSN: 2314-4912 (print), ISSN: 2314-4904 (online), Article ID: 537986

[15] Yaghoubi H. Application of magnetic levitation technology in personal transportation vehicles. Current Advances in Civil Engineering (CACE). 2013;**1**(1):7-11

[16] Yaghoubi H. Urban Transport Systems. University Campus STeP Ri, Rijeka, Croatia: InTech Publisher; 2017. ISBN: 978-953-51-2874-8, Print ISBN: 978-953-51-2873-1, 11 chapters. DOI: 10.5772/62814

The Feasibility of Constructing Super-Long-Span Bridges with New Materials in 2050

Faham Tahmasebinia,
Samad Mohammad Ebrahimzadeh Sepasgozar,
Hannah Blum, Kakarla Raghava Reddy,
Fernando Alonso-Marroquin, Qile Gao, Yang Hu,
Xu Wang and Zhongzheng Wang

Additional information is available at the end of the chapter

http://dx.doi.org/10.5772/intechopen.75298

Abstract

This chapter explores the possibility of designing and constructing a super-long-span bridge with new materials in 2050. The proposed bridge design has a total span of 4440 m with two 330-m end spans and a central span of 3780 m. The height of the two pylons is 702 m, and the deck width is 40 m. The features of this structure include the combination of a suspension bridge and cable-stayed bridge, application of carbon fibre materials, extension of deck width and pretension techniques. Linear static analysis, dynamic analysis and theoretical analysis are conducted under different loading cases. In linear static analysis, the stresses under critical load combinations are smaller than the ultimate strength of the materials. However, the maximum deflection under the dead and wind load combination exceeds the specified serviceability limit.

Keywords: long-span bridges, cables and tendons, materials technology

1. Introduction

The feasibility of designing a super-long-span bridge with new materials in 2050 is studied. Longer bridge spans have the benefits of increased horizontal navigation clearances and reduced risk of ship collisions with piers [1]. The length of very long span suspension and cable-stayed bridges are often limited by the weight of the cables. As spans increase, the cables experience high stresses due to their own self-weight, and the overall structure becomes less stiff as the stiffening contribution of the deck becomes negligible [1]. Therefore, strong but light-weight materials must be used in the design of super-long-span bridges.

There are many new high-strength materials with low density like carbon fibre with epoxy, graphene oxide and alumina-polymer composites. Some materials have much better mechanical performances than steel or concrete but are only used in some high-tech industries such as aerospace, wind energy and automotive industries due to their high price. By 2050, the new materials are likely to be used extensively in construction due to the reduced cost in the development process of new materials [2]. This paper presents and analyses a super-long-span bridge design which has a total span of 4440 m with 40 m width deck and two 702-m-high pylons. The bridge design is based on the Golden Gate Bridge and a finite element model is created in Strand7 [3] which is a modification to the Golden Gate Bridge model developed by [4]. The central span of the bridge is 3780 m, which is three times the span of the Golden Gate Bridge of 1260 m, while the length of the two end spans is the same at 330 m. Previous studies have been conducted on super-long cable-stayed bridges using carbon fibre reinforced polymer [5] and on long-span suspension bridges using fibre-reinforced polymer [6]. Special techniques are adopted in this design where the bridge combines the advantages of a suspension bridge and a cable-stayed bridge to minimise the deflection of the superstructure and the pylons. The material of the catenary cables and stay cables are changed to a lightweight fibre carbon composite [7] with high stiffness and high strength, and standard carbon fibre is used in the superstructure and the vertical hangers. Finally, the stayed-cables of the bridge are pre-strained in this design.

Carbon nanofibres have cylindrical shapes with graphene layers constructed in the morphology of cones or plates or sheets, with an average diameter of 50–100 nm and an average length of 50–200 um, exceptional thermal and mechanical properties (as high as elastic modulus of 600 GPa, tensile strength of 8.7 GPa, surface area 40 m^2/g), which offer a wide range of applications in the civil engineering discipline (e.g. bridges, roads, railways, tunnels, airports, ports and harbours), and other areas such as aerospace, automotive, sports goods material, and so on. Reinforcement of such new nanocarbons with polymeric materials further boosts their mechanical properties through different fabrication technologies such as wet/hand lay-up/spray lay-up, autoclave curing, filament winding, pultrusion, wet/hand lay-up, and so on. These extraordinary properties of advanced hybrid composites have enabled the design engineers to use them in the renewal of civil infrastructure ranging from the strengthening of reinforced concrete, steel and iron, and for replacement of bridge decks in rehabilitation (seismic repair, strengthen or retrofitting) to the construction of new ultra super-long bridge and building structures with less cost. In 1972, high strength polymeric material roof structure with the shape of an umbrella was manufactured via hand lay-up fabrication process and transported from the UK to be erected at the international airport of Dubai. In 1990s, it was replaced by advanced composites that were made with sophisticated glass fibre-reinforced plastics. Such advanced polymer composites with nanofillers (e.g., nanocarbons, glass fibre) are used for the development of building systems and building blocks using an automated construction system (ACS), which consists of a number of interlocking fibre-reinforced polymers (e.g., aramid) that can assemble into a large number of different efficient civil structures (e.g., 3D form) for use in the construction industry. Some examples of these ACS systems in the area of bridging engineering are Humber Bridge (1410 m span), Aberfeldy Bridge, Iron Bridge and the Bonds Mill Bridge from UK, and Gilman Bridge (450 m span), George Washington Bridge

(1067 m span), Golden Gate Bridge (1280 m span) and Wickwire Run bridge from USA, Forth Road Bridge (1005 m span) from Scotland, and Tagus Bridge (1013 m span) from Portugal, and so on. These ACS components are cheaper with durability, light weight, low cost, speed of construction, ease of transportation, and they show superior mechanical properties (e.g., tensile, compression, shear strength) than iron, steel and stone. Therefore, such advanced carbon fibre-reinforced polymer (CFRP) composite materials are promising candidates in the future for the construction of ultra-super-long bridges. It is feasible to develop over 10,000 span stable super-long bridges using new concepts. A new concept of engineering that is used for nanoscale modelling of super-long bridges can be described in following sections.

Finite element models were created in Strand7 [3], and the results show that a maximum deflection of 8.3 m occurs under the combination of dead and wind (G+W) load, which slightly exceeds the AS5100 [11] limit of 6.3 m. Furthermore, a 32.4 m transverse deflection is found under the dynamic wind analysis. Maximum tensile stresses of 1154 and 1152 MPa are observed in the catenary cables and stayed cables, respectively, which are below the tensile strength of 1600 MPa of M55**UD carbon fibre.

2. Structural system

2.1. Structural members

The design of this long-span bridge is based on the Golden Gate Bridge, therefore structural member types are essentially similar to the Golden Gate Bridge, which consists of a bridge deck with a supporting trusses and beams system, two pylons, catenary cables, vertical hangers and stayed cables and eight lanes of vehicle traffic. The design of the central span between the two pylons is based on a typical suspension bridge, while the two edge spans are similar to a cable-stayed bridge. The superstructure spanning between the two pylons is hung by vertical suspenders at 15-m intervals, which is the same as a typical suspension bridge. These vertical hangers carrying the loads on the deck are supported by the catenary cables suspended between the two pylons. Additionally, the stay cables at the two edge spans connecting the top of the pylons and the ends of the bridge are anchored by the abutment anchors at each end of the bridge. The cables directly running from the tower to the deck form a fan-like pattern on a series of parallel lines. Due to the different stress modes on the structural members, different materials are selected for each member based on their properties such as ultimate tensile and compressive strength, density and Young's modulus. The properties of the materials used in the bridge model are listed in **Table 1** [7–9]. The material selection is further discussed for each structural member.

The superstructure of the bridge consists of four major components: the bridge deck, permanent formwork, the cross girder and the deck truss system. The 0.5-m-thick bridge deck is made up of reinforced concrete while the material applied to the rest of the components of the superstructure is standard carbon fibre to reduce the self-weight of the superstructure. **Figure 1** shows the details of the arrangement of the structural members in the superstructure (without the bridge deck). As shown in **Figure 1**, the truss system resisting tensile or compressive force is attached to the cross girders running across the driving direction of the bridge.

Materials	Ultimate tensile/compressive strength (MPa)	Density (kg/m³)	Young's modulus (GPa)
Steel (Grade 350)	350 (C/T)	8000	207
Reinforced concrete	25.5 (C)	2500	34.5
Standard carbon fibre fabric	600 (T)	1600	70
Carbon fibre composite (M55** UD)	1600 (T)	1650	300 (0°)

Table 1. Properties of the material used in the model.

A 2.5 m-deep UB section is selected for the cross girders to carry flexural loads and then to transfer the loads into the truss system below. The concrete bridge deck sits on top of the truss system and cross girders, and the live load and vertical wind load are directly applied on the top of the bridge deck. Furthermore, the rectangular hollow section cross bracings distribute the loads on the deck onto the cross girders, and they also perform as the tensile reinforcement for the bridge deck above.

The design uses three types of cables: catenary cables, vertical hangers and stayed cables. In the central suspended deck, cables suspended via pylons hold up the road deck, and the weight and the vertical loads are transferred by the cables to the towers, which in turn transfer to the pylons and the anchorages at the end of the bridge. Since all of the cables are in tension, a lightweight carbon fibre or carbon fibre composite should be used in the cables based on material properties in **Table 1**.

Firstly, the catenary cables with a diameter of 2.2 m are suspended between the pylons, with a total cable length of 3915 m and further extend and transfer the loads to the anchorages at the bank. Carbon fibre composite (i.e., M55**UD) is used in the catenary cables. The shape of the catenary cables is determined by selecting an appropriate interpolated shape between the

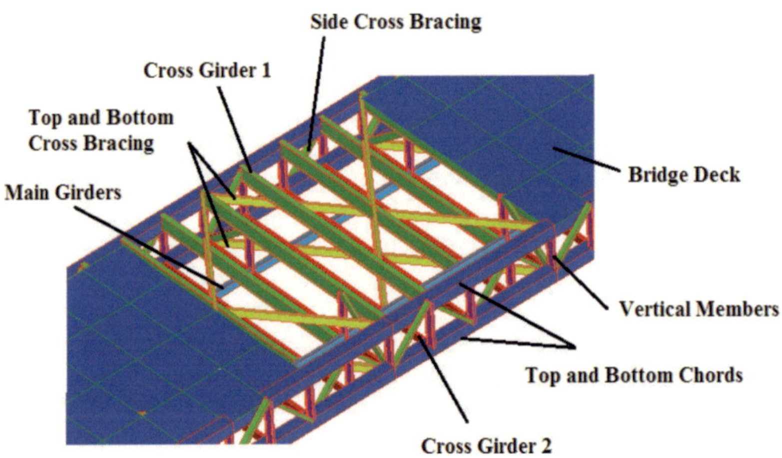

Figure 1. Details of the bridge superstructure.

catenary shape and the parabolic shape. The detailed explanation of this process is introduced in Section 3. The catenary cables are formed by connecting the coordinates that mimicked the shape of the cable, so the cables are segmented instead of smooth. Secondly, there are 252 pairs of vertical hangers at 15-m intervals at the central span. The diameter of the vertical hangers is 0.16 m. Standard carbon fibre is used in vertical hangers due to the relatively low tensile stress. Thirdly, the stayed cables at the two side spans are made up of M55**UD with a diameter of 0.15 m.

Two pylons are also built up at positions which are 330 m from each end of the bridge. The total height of a pylon is 702 m. The superstructure is connected to the pylons at 216 m from the foundations of the pylons. The loads on the catenary cables and the stayed cables are transferred to the pylons as a compressive force; therefore, Grade 350 steel is used as the material of the pylons [10].

The sizes and materials of the structural elements of the long-span bridge are shown in **Table 2**.

2.2. Structural system resisting vertical loads

The vertical loads acting on the bridge mainly consist of the self-weight of the structural members, the live load due to traffic and the vertical wind load. The vertical loads applied to the bridge deck are firstly carried by the reinforced concrete deck through bending, where the deck is directly supported every 5 m by the cross girders (I beams) with web stiffeners. The web stiffeners act to increase the shear capacity of the deep I beams and decrease the chance of shear

Structural elements details	Material	Suggested structural element sizes
Bridge deck	Reinforced concrete	0.5 m thick
Cross girder 1	Standard carbon fibre fabric	2500UB3650
Cross girder 2	Standard carbon fibre fabric	1200×500×50 RHS
Top and bottom chords (blue)	Standard carbon fibre fabric	2000×2000×200 SHS
Top/Bottom cross bracing (yellow)	Standard carbon fibre fabric	1200×500×50 RHS
Side cross bracing (green)	Standard carbon fibre fabric	1000×500×50 RHS
Main girders (blue)	Standard carbon fibre fabric	1400×700×75 RHS
Vertical members (pink)	Standard carbon fibre fabric	1500UB1420
Catenary cables	Carbon fibre composite (M55** UD)	2.2 m diameter
Vertical hangers	Standard carbon fibre fabric	0.16 m diameter
Stayed-cables	Carbon fibre composite (M55** UD)	0.15 m diameter
Pylons	Steel (Grade 350)	16,000×10,000
Pylon diagonal bracing	Steel (Grade 350)	5000×5000
Pylon cross bracing	Steel (Grade 350)	8000×16,000

Table 2. Structural elements for the super-long-span bridge.

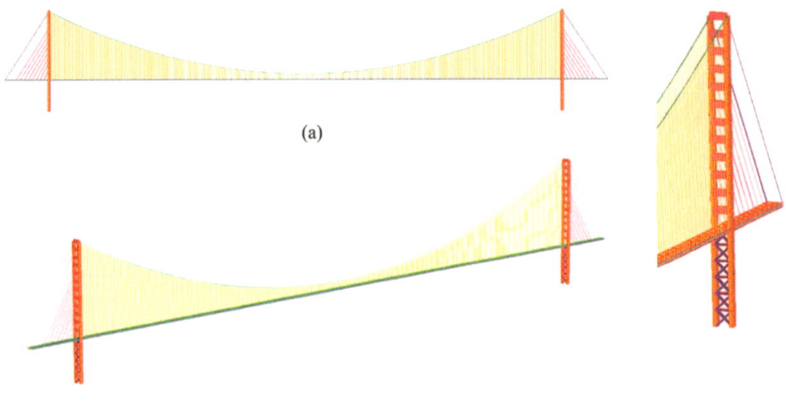

Figure 2. Side views and 3D view of the Strand7 model (a) Longitudinal view, (b) Details of Pylon, (c) 3D view.

buckling in the web. Then the cross girders transfer loads from the bridge deck to the truss system below the deck through bending. In the Strand7 model, the truss members are modelled as rectangular hollow sections to simplify the design. The members of the truss system can only carry axial force, so the top chords are in compression, and the bottom chords are in tension.

The loads on the truss system are spread along the main longitudinal truss members, and further transferred to the vertical hangers which are hanging off the corresponding superstructure every 15 m. These cables carry the loads from the bridge deck up to the catenary cables and the stayed cables through pure tension. On the Golden Gate Bridge, each catenary cable is made up of 27,572 galvanised steel cables which are grouped into 61 cable groups, which are then bunched together to form the 0.92 m diameter cable. For the super-long-span design, the larger diameter of catenary cables and stayed cables requires more galvanised steel cables to group larger cables. These stayed cables are also anchored at the abutments to keep them in tension and to pass the tensile load into the ground through the abutments. The pylons supporting the catenary cables, the stayed cables and the bridge deck are loaded in compression (**Figure 2**).

2.3. Structural system resisting lateral loads

Only wind load is considered as the lateral load acting on the bridge. Because this bridge is very long, the frequency of earthquakes is not consistent with the resonant frequency of the bridge. Therefore, the action of the earthquake load is not significant in this design. For simplicity, it is assumed that the transverse wind load only acts on the superstructure and the pylons. Therefore, the primary system used to resist transverse wind loads consists of the superstructure at the central span which is mainly restrained by the two pylons, the vertical hangers which are further suspended from the catenary cables and two pylons resisting the wind transverse wind load.

2.4. Loads

Dead load (G) accounts for the self-weight of the entire structure, which is calculated by multiplying member dimensions with the corresponding density. The Strand7 model calculated the structure's mass automatically after input of the material density and geometry. Dead load is then calculated by multiplying the density by the gravitational acceleration.

According to AS5100.2 Bridge Design Codes [11], the live load (Q) applying on the bridge deck is the load resulting from the passage of vehicles and pedestrians, which is SM1600 loading. However, for simplicity, the live load is considered as a pressure acting on the bridge deck in the Strand7 model. The most severe load specified in AS5100.2 is added together and averaged as a pressure load over the bridge deck, resulting in a surface pressure of 10.416 kPa. It is recommended that the Load Influence solver in Strand7 can be used to determine the critical point for the live load and the sensitivity of structure members.

Wind load is the dominant impact on super-long bridges. The first step is to obtain the design wind speed calculated as specified in AS/NZS1170.2 [12], which is derived from regional basic wind speed after adjustment for average return interval, geographical location, terrain category and height above ground. The site wind speed is calculated as:

$$V_{sit,\beta} = V_R M_d (M_{z,cat} M_s M_t) \tag{1}$$

Since most of the factors vary with different site conditions, and the location of the bridge is not determined yet, by looking at a bridge over Bemboka River in New South Wales, a serviceability design wind speed V_s=37 m/s and an ultimate design wind speed V_u=48 m/s are used in the design. According to AS5100.2 [11], the ultimate design transverse wind load and ultimate vertical wind load are

$$W_t^* = 0.0006\, V_u^2 A_t C_d = 1.935\, A_t \tag{2}$$

$$W_v^* = 0.0006\, V_u^2 A_p C_L = 1.037\, A_t \tag{3}$$

where A_t and A_p are the bridge area in plan, C_d is the drag coefficient, and $C_L = 0.75$ is the lift coefficient. Then the wind load applied on each structure member can be calculated for each structural member. For simplicity, the transverse wind load is assumed to only act on the superstructure and the pylons.

AS5100.2 Clause 23 [11] specifies that for *Ultimate Limit State (ULS)* analysis, the load combinations should include Permanent Effect + Road/rail traffic load and Permanent Effect + Wind load. Therefore, in the static analysis and dynamic analysis, the load combinations of G+Q and G+W are considered.

3. Theoretical analysis

By applying fundamental principles in engineering design, analytical calculations were carried out to determine the optimum cable shape for the suspension bridge and to predict and verify the maximum stresses in the catenary cables and the natural frequency of the structure.

3.1. Optimum cable shape

The shape of a flexible cable under self-weight is a catenary [13]. The equation for a catenary is:

$$y_c = a \times \cosh\left(\frac{x}{a}\right) \tag{4}$$

where the parameter a can be determined once the points through the catenary are known.

Alternatively, when the cables are under heavy load (i.e. the self-weight of cables is negligible compared to the applied uniformly distributed load), then the shape becomes a parabola with the equation:

$$y_p = \frac{w}{2T_0}x^2 + \beta \tag{5}$$

where T_0 is the tensile force in the middle of the cable, w is the uniformly distributed load, and β is the distance between the lowest point of the cable and the top surface of the deck. However, in this design of a super-long-span suspension bridge, neither the self-weight of cables nor the applied uniformly distributed load from the deck can be ignored. Therefore, the resulting shape of the catenary cable is between the shape of a parabola and a catenary [14]. To determine the optimum cable shape that results in minimum deflection for the suspension bridge, an interpolation factor, K, is introduced to determine the final cable shape:

$$y = K \times y_c + (1 - K) \times y_p \tag{6}$$

where $K \in [0, 1]$. Note that the cable shape will be a catenary when K is 1 or a parabola when K is 0.

When only considering dead loads, the normalised maximum deflections of the middle span are plotted against different K values as shown in **Figure 3**. The normalised deflection is equal

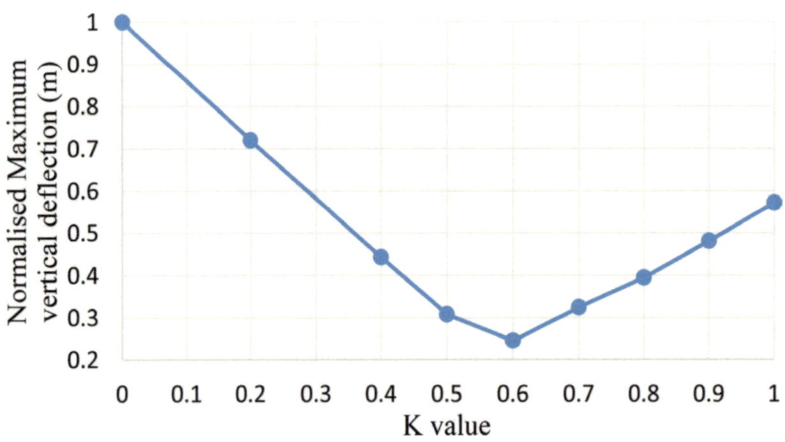

Figure 3. Normalised maximum deflection under self-weight vs. K.

to the maximum defection with a specific K value divided by the maximum deflection with K = 0. It is observed that for different K values, the maximum deflections for each case are different, and it reaches a minimum when K is about 0.6. It is also important to note that when the applied load has changed, the optimum K value will change as well. For example, when considering the live load by applying additional uniformly distributed load onto the deck, the optimum value K will increase since the overall load will be closer to a uniformly distributed load where K = 1 (i.e. a parabolic shape will have minimum deflections), so the optimum cable shape would be closer to a parabola. By considering different loading cases while maintaining relatively low deflections, a K value of 0.7 was adopted for the design.

3.2. Maximum tensile stress in main span cables under dead loads

To determine the maximum tensile stress in cables, four assumptions are made to simplify calculations. First, the combination of the self-weights of the cables and the decks is regarded as a uniformly distributed load, which means the consequent cable shape is parabolic. Second, the displacement of the top point of the pylons can be neglected by assuming pin connections between cables and pylons. Third, the deflection of cables under loading is not significant meaning that it will not affect the loads. Fourth, the influence of earth curvature is ignored, which means the gravitational forces are perfectly downwards. The free body diagram of half internal span is shown in **Figure 4**.

From **Figure 4**, T_0 is the horizontal tensile force at the midpoint, w is the uniformly distributed load, L is the total length of internal span, h is the extreme difference of the main cable in the vertical direction and T_{max} is tensile stress at the pylon.

By applying the moment equilibrium at the top right corner, it can be obtained that:

$$\sum M = T_0 \times h - w \times \frac{L}{2} \times \frac{L}{4} = 0 \rightarrow T_0 = \frac{wL^2}{8h} \quad (7)$$

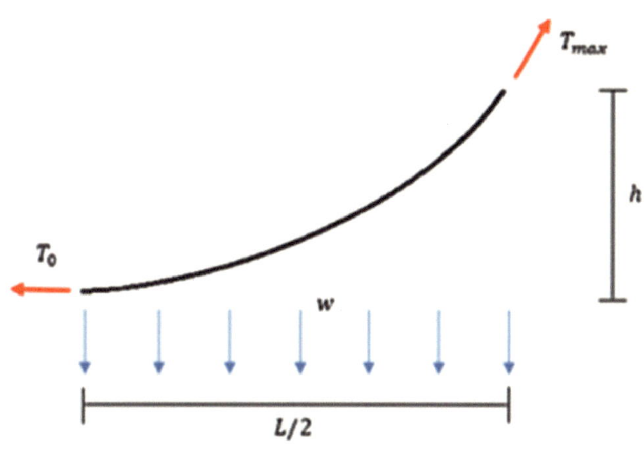

Figure 4. Free body diagram of main cable—Half internal span.

Then, by applying Pythagorean theorem, the T_{max} is obtained:

$$T_{max} = \sqrt{T_0^2 + \left(\frac{wL}{2}\right)^2} = \sqrt{\frac{w^2 L^4}{64 h^2} + \frac{w^2 L^2}{4}} \tag{8}$$

From Strand7, the dead load of the deck and cables is obtained: $w = 485.275\, kN/m$. With $L = 3780$ m, $h = 443.5$ m, the horizontal tensile force is calculated to be $T_0 = 1954.29$ kN, and $T_{max} = 2158.80$ kN. Then, for a cable with diameter D = 2.2 m, the stresses are calculated by dividing the force by the cross-sectional area of cable: $\sigma_0 = \frac{T_0}{A} = 515$ MPa, and $\sigma_{max} = \frac{T_{max}}{A} = 568$ MPa.

3.3. Bridge natural frequency

The natural frequency of bridges is affected by different material and geometric properties:

$$n = f(E, I, \rho, k, L) \tag{9}$$

where E is elastic modulus, I is the second moment of area, ρ is the density of materials, k is the factor depending on the boundary condition and L is the span distance of the bridge.

Since the equation for the natural frequency for a complete representation of the suspension bridge is extremely complicated, a simplified method was adopted under the following three assumptions. First, the oscillation mode will be such that the bridge will deflect in the transverse direction, which is consistent with the first mode determined from the numerical analysis presented. Second, the geometric structure of the deck will be regarded as a simple beam. Third, the cables provide no significant effect on the oscillation frequency in the transverse direction. By applying a simplified equation:

$$n = \frac{K}{2\pi L^2}\sqrt{\frac{EI}{m}} \tag{10}$$

where K is a constant depending on the mode of vibration mode, m is the mass per metre for the deck, then the natural frequency of the first mode is determined to be 0.0212 Hz.

4. Numerical analysis

A detailed finite element model of the structure was created in Strand7 [3], and details of the model are described.

4.1. Mesh and mesh quality

In Strand7, plate elements are used to model the bridge concrete deck, while the remainder of bridge elements use cut-off bars and beam elements. Only the plate elements within the model need meshing as cut-off bars and beam elements in the truss system do not require meshing. However, the node positions of truss members below the bridge deck have already subdivided the deck plate, resulting in a relatively coarse mesh. Since the bridge deck is not a critical component in the analysis process, the mesh quality of plate elements is not an issue.

4.2. Element types

For the bridge superstructure, the bridge lanes are modelled by plate elements having concrete material properties. The components of the truss system supporting the concrete deck are modelled as beam elements in the shape of I sections or rectangular hollow sections (RHS). In addition, the pylons, pylon diagonal bracings and pylon cross bracings are modelled as B2 elements (beam element) with solid rectangular sections.

The catenary cable is modelled using cut-off bar elements which are connected between nodes with coordinates that are determined by a theoretical analysis using both catenary curve and parabolic curve equations. The optimum cable shape results in an acceptable maximum deflection. The nodes for the catenary cables were imported into the Strand7 model, and were then joined by B2 cut-off bar elements to form the catenary cables. The vertical cables are also modelled by connecting the nodes on the catenary cables and bridge deck by using cut-off bar elements. Similarly, the stay cables are also modelled by connecting the nodes on the pylons and edge span bridge decks with cut-off bars. The reason for selecting cut-off bar is that this type of beam element only allows axial forces and allows users to set the tensile or compressive capacity of the bar. Therefore, all of the cables can be set to have zero compressive capacities.

The nodes on the base of the pylons were completely fixed in all six degrees of freedom, which includes DX, DY and DZ as well as RX, RY and RZ, which is consistent with real structural behaviours with a deep and strong foundation system. Meanwhile, four giant anchors located at the ends of the bridge span are also fully fixed to prevent bridge movements in all directions.

4.3. Solvers used in the analysis

Both the static solver and the dynamic solver are used in the numerical analysis. The deflection of the bridge and the stress within the structure members under dead load, live load, wind load and their combinations are determined by the linear static solver. The non-linear static solver is used to guarantee the maximum stresses within the members are below the yield stress of the corresponding materials.

The Natural Frequency Solver is used to determine the natural frequencies of the bridge based on the results of linear static analysis. Linear static analysis under the G+Q load is conducted first since the tension and compression within structural members would have significant impacts on its natural frequencies. Up to 50 frequency modes are analysed in order to obtain the sufficient mass participation of the bridge.

Once the natural frequency analysis is completed, the harmonic response analysis is performed to determine the mass participation of the bridge and its displacements under wind load. Case factor is set as 1.0 for wind load only, a 5% modal damping is applied in harmonic analysis and all 50 natural frequencies are investigated. The resonance frequency and the corresponding deflection can be obtained in the harmonic response analysis.

Spectral response solver is used to investigate the dynamic response to wind load, with the aid of a power spectral density curve. A 5% modal damping is applied and the results are calculated based on square root of the sum of the squares (SRSS) approach.

5. Structural design

In terms of structural design of the bridge, the capacity is defined as the strength limit state by the critical members which experience maximum axial force, bending moment or shear force. The serviceability limit state is also considered; therefore, the deflections of the deck are checked. In particular, carbon fibre has no yielding behaviours, so there will be no sign of failure in the carbon fibre cables. For members consisting of standard carbon fibre and M55**UD carbon fibre composite, the tensile strength is reduced by 80% due to this brittle nature. For steel members, the section yield stress is factored down to 90% of full yielding. According to the Steel Structures design code AS4100 [8], the nominal section capacity for members subject to axial tension and compression is calculated using the equations:

$$f_t^* = 0.85\, \varnothing\, k_t f_u \tag{11}$$

$$f_c^* = \varnothing\, k_t f_y \tag{12}$$

Assuming the tensile load is distributed uniformly to the catenary cable, the value of k_t equals to 1. As the catenary cables made of M55**UD carbon fibre composite are only subjected to tensile forces, the factored tensile capacity f_{t_cc} can be obtained as:

$$f_{t\_cc} = 0.85 \times 0.9 \times 1600 = 1224 \text{ MPa} \tag{13}$$

Similarly, for the truss members and vertical cables which are made up of standard carbon fibre, it is assumed that their effective cross-sectional area equals their total cross-sectional area ($k_f = 1$). Therefore, the factored tensile capacity f_{t_c} and compressive capacity f_{c_c} can be obtained as:

$$f_{t\_c} = 0.85 \times 0.9 \times 600 = 459 \text{ MPa} \tag{14}$$

$$f_{c\_c} = 0.9 \times 570 = 513 \text{ MPa} \tag{15}$$

Pylons are made up of Grade 350 steel, where the factored yield capacity f_{c_p} is calculated as:

$$f_{c\_p} = 0.8 \times 350 = 315 \text{ MPa} \tag{16}$$

5.1. Cable design

For bridges with super-long spans, the most challenging and critical members are vertical cables and catenary cables because the dead and live loads acting on the bridge deck are mainly transferred to the vertical cables and further to the catenary cables, which result in large tensile stresses in those cables. As specified above, the vertical cables and catenary cables are cut-off bars, which only resist axial tensile force. The Strand7 linear static analysis results under the G + Q case give the maximum tensile stresses in the cables compared to the capacities in **Table 3**.

For catenary cables, the maximum tensile stress is applied at the end segments near the pylons, which is reasonable since loads on the catenary cables are transferred to the pylons and further to the stayed cables. The last column in **Table 3** is the percentage of maximum applied stress to the tensile capacity. The results show that the current cable design is optimistic as the applied stresses are extremely close to the cable capacity, and the maximum bridge deflection under G+Q is relatively close to the AS5100 [11] limit. However, it is significant to note that the dimension of the catenary cable is relatively large compared to the original design, as it was increased from 1.20 to 2.20 m in diameter. To prove the validity of the Strand7 model, the maximum applied stress in the catenary cables in Strand7 analysis is compared with the analytical results in later sections.

5.2. Deck design

In terms of the deck design, the deck members are divided into two groups: truss members and flexural members. Truss members such as the top and bottom chords resist axial load only, so the tensile and compressive axial strength for each type is compared with the maximum applied axial loads under the most critical load combination (G+Q). Flexural members such as the top and bottom cross girders also experience significant bending moments; thus, the calculated moment section capacity for each type of member is further compared with the maximum applied bending moments under the G+Q case. In particular, for a conservative design, the axial and bending capacities in **Table 4** are factored into capacities.

The last column in **Table 4** indicates the larger percentages of applied axial stress to the axial capacity of critical axial force in terms of tension or compression. All truss members have a

Member type	Member dimensions (m in diameter)	Tensile capacity (MPa)	Maximum stress (MPa)	Capacity (%)
Vertical cables	0.16	459	416	90.63
Catenary cables	2.20	1224	1154	94.28
Stayed cables	0.15	1224	1152	94.11

Table 3. Comparison of tensile capacity to maximum applied stress (suggested design).

Truss members	Geometry	Capacity		Loading		Capacity (%)
		f_{u_t} (MPa)	f_{u_c} (MPa)	Design tension (MPa)	Design compression (MPa)	
Bottom cross girders	1200×500×50 RHS	459	513	53	8	11.5
Top/bottom chords	2000×2000×200 RHS	459	513	72	32	15.7
Top/bottom cross bracings	1200×500×50 RHS	459	513	29	15	6.3
Side cross bracings	1000×500×50 RHS	459	513	208	285	55.6
Main girders	1400×700×75 RHS	459	513	10	10	2.2
Vertical bracings	500UB667	459	513	66	17	14.4
Top cross girders	2500UB3650	459	513	10	19	3.7

Table 4. Design of deck truss members based on axial capacity (current design).

percentage of capacity less than 100%, which indicates that this design satisfies the strength limit state requirements. However, except for side cross-bracing members, the percentages of capacity for other truss members are less than 15%, which means that these members are over-conservative. Hence, the geometries for those members can be adjusted to achieve a more economical design.

For the deck flexural members under bending, the section capacity can be obtained by:

$$M = \frac{\sigma y}{I} \tag{17}$$

In terms of the section modulus and the axial compression capacity, the section capacity for the top and bottom cross girder and top and bottom chords is calculated as follows (**Table 5**):

$$M_{t\_g} = 0.262 \times 459 \times 10^3 = 1.203 \times 10^5 \text{ kNm} \tag{18}$$

$$M_c = 0.787 \times 459 \times 10^3 = 3.612 \times 10^5 \text{ kNm} \tag{19}$$

Flexural members	Geometry				Capacity		Loading	Capacity (%)
	Area (m²)	Ix (m⁴)	y (m)	Z_x (m³)	f_{u_c} (MPa)	Section capacity (kNm)	Design moment (kNm)	
Top cross girder	0.465	0.328	1.25	0.262	459	1.203×10^5	12,025	10
Top/bottom chords	1.440	0.787	1.00	0.787	459	3.612×10^5	315,000	87.2
Bottom cross girder	0.160	0.028	0.60	0.046	459	0.21×10^5	725	3.5

Table 5. Design of deck flexural members based on moment section capacity (current design).

$$M_{b\_g} = 0.046 \times 459 \times 10^3 = 0.21 \times 10^5 \text{ kNm} \tag{20}$$

Similarly, all deck flexural members have a percentage of capacity less than 100% while for the top and bottom cross girders the value is below 15%, so the member geometries are necessarily optimised to achieve a more economical design.

6. Results and discussion

6.1. Static analysis

Using the linear static solver available in Strand7, the suspension bridge is assessed in terms of deflections and stresses under the G+Q and G+W cases. As specified in AS5100.2 [11], the ratio of the allowable vertical deflection or transverse deflection over the length of the span is $\frac{1}{600}$, which corresponds to a maximum deflection of 6.3 m. The allowable deflections are further compared with the results in Strand7 analysis.

6.1.1. Dead load + live load

The deflection contour of the suspension bridge under the combination of dead and live loads is shown in **Figure 5**. The maximum vertical deflection occurs at the middle of the central span, which is around 8.3 m downwards. The vertical deflection exceeds the allowable deflection limit, which is 6.3 m, hence more techniques are required to further decrease the vertical deflection.

In terms of the stresses shown in **Figure 6**, the maximum stresses occur in the catenary cables and stay cables, with the values of 1154 and 1152 MPa respectively. As calculated in Section 5, the carbon fibre composite (M55**UD) used in catenary and stay cables have a factored tensile capacity of 1224 MPa and can sustain the load without failure. Additionally, the vertical cables within the central span have relatively small stress at around 400 MPa, which is below the factored tensile capacity of 459 MPa.

Figure 5. Vertical deflection contour under dead + live load.

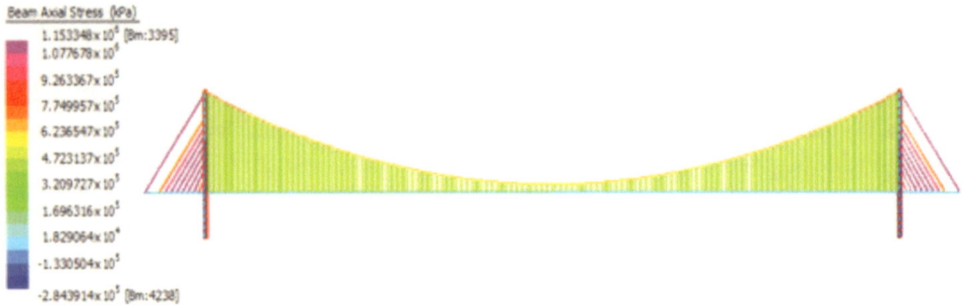

Figure 6. Beam axial stress (kPa) contour under dead and live load combination (G+Q).

6.1.2. Dead load + wind load

A horizontal wind load is applied onto the deck system and pylons. Under linear static analysis, the transverse deflection contour of the "Dead +Wind" loading case is shown in **Figure 7**.

It is found that the transverse deflections at the edge span and the central span near the pylons are very small, and the maximum transverse deflection of 30.9 m occurs at the mid-span of the central span. The transverse deflection under the "Dead load + Wind load" case exceeds the allowable value of 6.3 m, since the out of plane stiffness of the deck is insufficient via the lateral wind load. Therefore, the deflection under the "Dead load + Wind load" case is not satisfactory, and further methods should be considered to reduce the transverse deflection such as improving the stiffness of the deck.

In terms of the stress contours shown in **Figure 8**, the maximum stresses in this case are much lower than the stresses under the "Dead load + Live load" case which indicate that the "Dead load + Wind load" case is not the most critical case for element stresses. Hence the structural elements sufficiently sustain the loads without failure under the "Dead load + Wind load" case.

The catenary cable stress at the mid-span from the Strand7 analysis is compared to the analytical results. Under the "Dead load + Live load" case, the maximum catenary cable stress at the mid-span is 714 MPa, which is located at the region near the pylons, and the horizontal stress at the midpoint of the central span is 646 MPa from Strand7 linear static analysis. **Table 6** shows that there is a 20% difference between the numerical and analytical results.

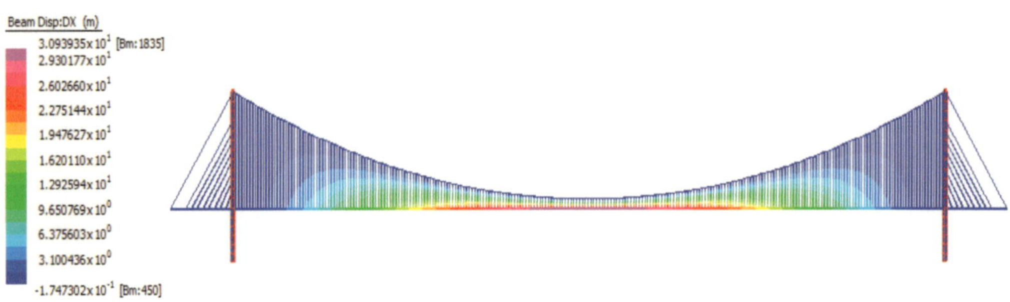

Figure 7. Transverse deflection contour under dead load and wind load.

http://dx.doi.org/10.5772/intechopen.75298

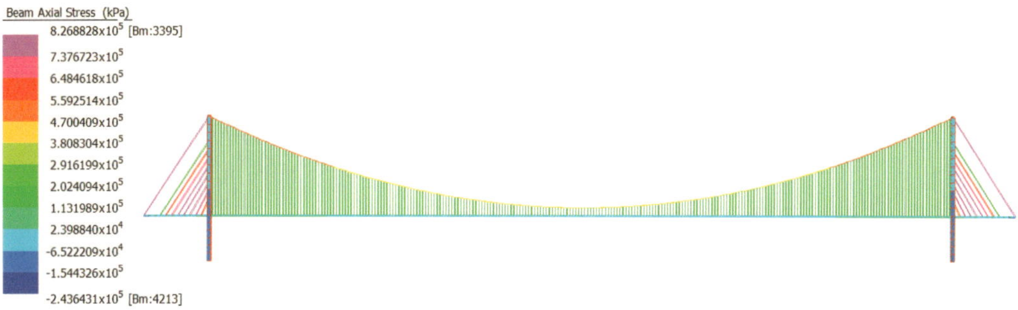

Figure 8. Beam axial stress (kPa) contour under dead and wind load combination (G+W).

6.2. Natural frequency analysis

Natural frequencies of the bridge are solved by Strand7 based on a linear static analysis result under the "Dead load + Live load" case since wind impact is significant when the bridge is in service. Up to 50 different modes of natural frequencies are solved.

In the first mode with a frequency of 0.00338 Hz, the bridge deforms horizontally in a single curvature shape. In the second mode, the bridge deforms horizontally in a double curvature shape, where the frequency is 0.06001 Hz, and in the third mode, the bridge deforms vertically in a double curvature shape, where the frequency is 0.071586 Hz. Further natural frequency modes with higher frequencies are not considered as critical as the first three modes, but the harmonic response solver will be used to investigate the wind impact of these modes to find the most disastrous wind speed.

6.3. Harmonic response analysis

Based on the previous solved natural frequencies, the harmonic response analysis runs under the "wind load only" case with 5% modal damping. **Figure 9** shows that the first mode is the most critical case, since at 0.00338 Hz, the bridge is under resonance and there is a relatively large transverse displacement of 47.6 m at the mid-span of the bridge. Specifically, the wind load is applied as steady sinusoidal force at various frequencies, where each cycle of loading exerts additional energy and increases the vibration amplitude. In this case, when the wind blows in the same frequency as the first natural frequency mode (0.00338 Hz), the bridge is under resonance where the maximum response amplitude occurs and the transverse displacement reaches 47.6 m. Therefore, the super-long bridge is considered to fail when the wind frequency coincides with the first mode of frequency of the bridge. However, the real behaviour of wind loading is further explored in Section 6.4.

	Numerical	Analytical	Difference (%)
Tensile stress near pylons (MPa)	714	568	20.4
Tensile stress at midpoint (MPa)	646	515	20.3

Table 6. Comparison of numerical and analytical catenary cable stresses.

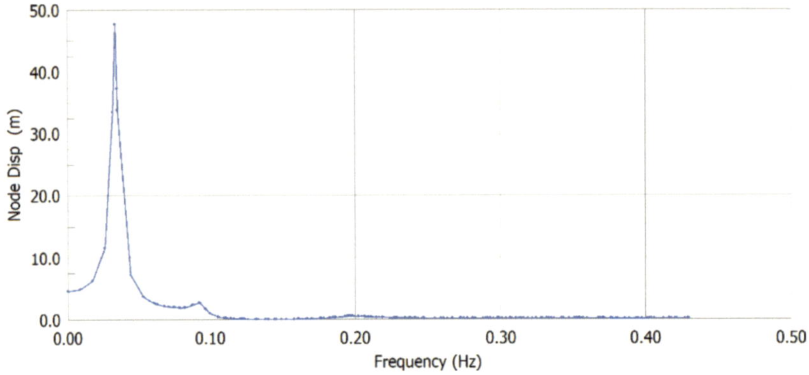

Figure 9. Frequencies vs. nodal displacement (D_x) at the mid-span.

6.4. Spectral response analysis

The linear elastic response of a super-long bridge subjected to a particular wind loading in Australia can be assessed by the "Spectral Response Solver" in Strand7, where the input for dynamic wind analysis can be expressed in terms of Power Spectral Density (PSD) curves [3]. The PSD curve (Davenport's equation) can be expressed as:

$$S_u(n) = 4\,k\,V^2\,\frac{x^2}{n\,(1+x^2)^{\frac{4}{3}}} \tag{21}$$

where $S_u(n)$ is the wind speed power spectral density, n is the frequency, V is the hourly mean wind speed at a 10-m height from the ground, $x = \frac{1200\,n}{V}$ and k is the roughness parameter. As a result, the equation for the Wind Force PSD is plotted as the factor vs. frequency graph in Strand7 and shown as **Figure 10**.

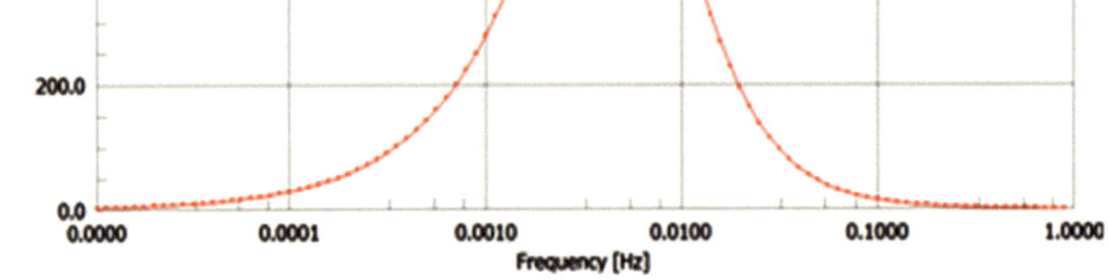

Figure 10. Factor vs. frequency graph for spectral response wind analysis.

http://dx.doi.org/10.5772/intechopen.75298

Figure 10 indicates that the wind load factor changes with wind frequencies and the largest wind load factor occurs when the wind frequency is approximately 0.0045 Hz. Based on the factor vs. frequency graph, the natural frequency of the bridge and applied wind pressure, the "Spectral Response Solver" is run and the result is calculated by the square root of the sum of the squares (SRSS) method in which the coupling among different modes is neglected. In addition, 5% modal damping is applied in the analysis. The result of transverse displacement is shown in **Figure 11**.

The result shows that a maximum transverse deflection of 32.4 m occurs at the mid-span, which is greater than the transverse deflection of 30.9 m obtained from linear static analysis. This phenomenon is reasonable because when the bridge natural frequencies coincide with the heavily factored wind frequencies, the wind impact is severely magnified. As the 32.4 m transverse deflection exceeds the AS5100 limit of 6.3 m, the serviceability criterion is not satisfied. The reason is that the out of plane stiffness of the bridge is insufficient to resist wind forces in this super long bridge. Therefore, further approaches are necessary to increase the lateral stiffness such as increasing the deck width and increasing the depth of the truss system.

6.5. Carbon fibre cost

Standard and ultra-high stiffness carbon fibres are mainly made for high-tech industries like the aerospace industry. They are very expensive and used in specialised applications such as aerofoils. The price for the ultra-high stiffness is \$2000 USD per kg [15], whereas standard carbon fibre was \$22 USD per kg in 2013 [15]. Currently, ultra-high stiffness is not economical for civil infrastructure. However, **Figure 12** shows the forecast cost of standard carbon fibre. The cost in 2017 is approximately \$12 USD per kg, and it has dropped significantly compared to the price in 2013. **Figure 12** shows that the decreasing price trend could be approximated to an exponential function as:

$$cost = k \times e^{-0.187 \times year} \quad (22)$$

where k depends on the year of interest. Based on the same trend, in 2050, the cost of standard carbon fibre would be \$0.02 USD per kg and the price of carbon fibre composite would be \$1.26 USD per kg, which are even cheaper than the price of steel now of \$1.60 USD per kg

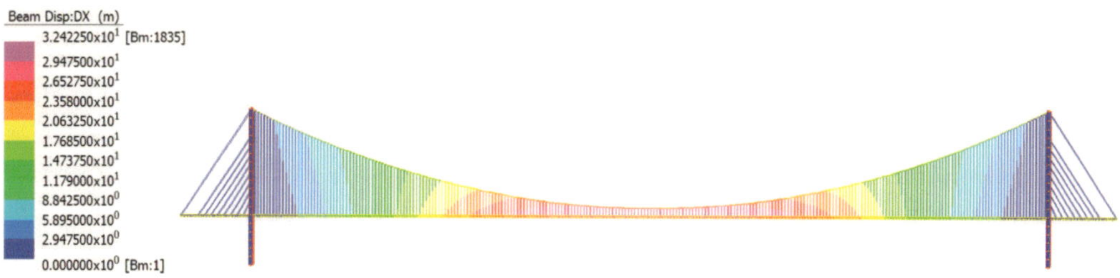

Figure 11. Transverse deflection contour under wind load (spectral response analysis).

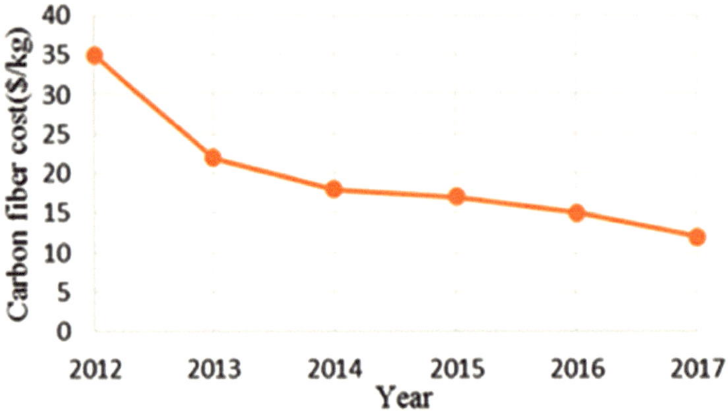

Figure 12. Carbon fibre cost forecast (USD).

[15]. Therefore, carbon fibre could be widely used in civil construction, and this design of a super-long-span bridge may be cost-effective to construct in 2050.

7. Conclusions

The proposed super-long-span bridge design has a total span of 4440 m with two 330-m-long end spans and a central span of 3780 m. Each pylon is designed to be 702 m high, and deck width is 40 m.

M55**UD carbon fibre is adopted for catenary cables and stayed cables because it has higher tensile strength and elastic modulus compared to steel. Standard carbon fibre with a high strength-to-weight ratio is used in the deck system and vertical cables to achieve a lighter self-weight and higher strength capacity. A new iterative technique is introduced and developed to determine the optimum cable shape to minimise the deflections by introducing a K factor. Another innovative design technique, the combination of a cable-stayed and suspension structure, is adopted to balance the deflections of the end and the central bridge spans. A pretension technique is also used in the stayed-cable design.

Overall, in the static analysis, the stresses found under the G+Q and G+W load cases are lower than the capacities of the materials, and the strength requirements are satisfied. However, the maximum deflections under static and dynamic analysis do not meet the criteria for the AS5100 limit of 6 m displacement, with an 8.3 m vertical deflection under the G+Q loading case, a 30.9 m transverse displacement under the G + W static analysis, and a 32.4 m transverse displacement observed in the wind dynamic analysis. Although standard carbon fibre and carbon fibre composite are currently very expensive, the price is expected to drop significantly by 2050 based on recent trends.

Further research is recommended to reduce the transverse deflection by considering increasing the lateral stiffness of the bridge. Additionally, the development of finite element models with more optimised structural members, section sizes and geometries are recommended to reduce the vertical deflections as well as the total cost of the bridge.

Author details

Faham Tahmasebinia[1,2]*, Samad Mohammad Ebrahimzadeh Sepasgozar[3], Hannah Blum[1], Kakarla Raghava Reddy[1], Fernando Alonso-Marroquin[1], Qile Gao[1], Yang Hu[1], Xu Wang[1] and Zhongzheng Wang[1]

*Address all correspondence to: faham.tahmasebinia@sydney.edu.au

1 School of Civil Engineering, The University of Sydney, Sydney, NSW, Australia

2 School of Mining Engineering, The University of New South Wales, Sydney, NSW, Australia

3 Faculty of Built Environment, The University of New South Wales, Sydney, NSW, Australia

References

[1] Clemente P, Nicolosi G, Raithel A. Preliminary design of very long-span suspension bridges. Engineering Structures. 2000;**22**(12):1699-1706

[2] Shama Rao N, Simha TGA, Rao KP, Ravi Kumar GVV. Carbon Composites are Becoming Competitive and Cost Effective. 2014. White paper from www.infosys.com

[3] Strand7 v2.4.6. Sydney, Australia: Strand7 Pty Limited; 2015

[4] Game T, Vos C, Morshedi R, Gratton R, Alonso-Marroquin F, Tahmasebinia F. Full dynamic model of Golden Gate Bridge. In: AIP Conference Proceedings, Vol. 1762(1). AIP Publishing; 2016, August. p. 020005

[5] Zhang XJ. Mechanics feasibility of using CFRP cables in super long-span cable-stayed bridges. Structural Engineering and Mechanics. 2008;**29**(5):567-579

[6] Yang Y, Wang X, Wu Z. Evaluation of the static and dynamic behaviours of long-span suspension bridges with FRP cables. Journal of Bridge Engineering. 2016;**21**(12):06016008

[7] ACP Composites Inc. Mechanical Properties of Carbon Fiber Composite Materials, Fiber/Epoxy resin (120°C Cure). [Online]. 2010. Available at: https://www.acpsales.com/upload/Mechanical-Properties-of-Carbon-Fiber-Composite-Materials.pdf [Accessed: 29 June 2017]

[8] Standards Australia. S4100.2:2016 Steel Structure. Sydney, Australia: Standards Australia; 2016

[9] Bluescope Steel. HA350 steel Datasheet Datasheets. 2017. See http://steelproducts.bluescopesteel.com.au/category/datasheets [Accessed: 29 June 2017]

[10] Grässel O, Krüger L, Frommeyer G, Meyer LW. High strength Fe–Mn–(Al, Si) TRIP/TWIP steels development—Properties—Application. International Journal of Plasticity. 2000;**16**(10):1391-1409

[11] Standards Australia. AS5100.2:2017 Bridge Design, Part 2: Design Loads. Sydney, Australia: Standards Australia; 2017

[12] Standards Australia. AS/NZS1170.2:2011 Structural Design Actions: Wind Actions. Sydney, Australia: Standards Australia; 2011

[13] Byer O, Lazebnik F, Smeltzer DL. Methods for Euclidean Geometry. U.S.A: MAA; 2010

[14] Kunkel P. Hanging With Galileo. Whistler Alley Mathematics [Online]. 2006. Available at: http://whistleralley.com/hanging/hanging.htm. [Accessed: 27 March 2016]

[15] Prince Engineering. Carbon Fiber used in Fiber Reinforced Plastic (FRP). [Online]. 2013. Available at: http://www.build-on-prince.com/carbon-fiber.html#sthash.gBoq2sJ6.dpbs. [Accessed: 3 July 2017]

Chapter 3

Developing a Bridge Condition Rating Model Based on Limited Number of Data Sets

Khairullah Yusuf and Roszilah Hamid

Additional information is available at the end of the chapter

http://dx.doi.org/10.5772/intechopen.71556

Abstract

This chapter utilizes artificial neural network (ANN) and multiple regression analysis (MRA) to model bridge condition rating based on limited number of data sets. Since data sets are very limited and there is a gap in range of rating scale, two conditions of data sets are used in this study, namely complete data sets and data set with bridge component condition rating data are missing. Five methods are then used to handle the missing bridge component condition rating data. Three commonly used methods and two new methods are explored in this study. It seems that the performance of the model using data sets after handling missing bridge component data to fill the gaps in the range scales of the bridge condition rating improved the performance of the model. In addition, a handling method that substitutes missing data of bridge component ratings with available bridge rating data is favorable. Based on the values of root mean square error (RMSE) and R^2, the ANN models perform slightly better than MRA to map relationship between bridge components and bridge condition rating. This concluded that ANN is suitable to model bridge condition rating compare to MRA method.

Keywords: multiple regression analysis, neural network, function approximation, bridge condition rating, root mean squared error

1. Introduction

The bridge condition rating is the most important part of the bridge management system (BMS) because, historically, bridge condition rating data were found to affect approximately 60% of the BMS analysis modules [1]. For BMS requirements, bridge inspectors generally use visual inspection as the first step toward condition assessment procedure unless a structure cannot be visually assessed. In through visual inspection, bridge inspectors evaluate the condition of a bridge using their personal experience and following guidelines such as those found in inspector's manuals. The visual inspection incorporates many parameters and human judgments that may produce slightly uncertain and imprecise results [2]. It would be ideal to conduct physical structural tests on each bridge component, but it would be impractical and economically prohibitive to implement them, given the large number of bridges to be inspected within a given period. As an alternative, the collective judgment of the inspectors can be used to develop unified, coherent bridge inspection procedures [3]. Thus, a mathematical method such as an artificial neural network (ANN) and multiple regression analysis (MRA) can be useful to handle this uncertainty, imprecision, and subjective judgment.

The limited inspection reports maintained by the Public Works Department (PWD) of Malaysia are used as the initial data, which contain valuable information about the condition ratings of the bridge components and whole bridges. The condition rating is a numerical system, where a number from one to five is assigned to each component of the structure based upon observed material defects and the resulting effect on the ability of the component to perform its function in the structural system, as described in Ref. [4]. **Table 1** shows the condition rating system used by PWD Malaysia. The bridge components that are inspected and rated contribute to the overall bridge rating, as shown in **Table 1**. The bridge components involved are the beam/girder, deck slab, pier, abutment, bearing, drainpipe, parapet surfacing, expansion joint, and slope protection.

The original data sets are very limited and many shortcomings exist in the records such as unavailable or missing bridge component condition rating data, the whole bridge condition rating is not distributed in complete range of rating scale and many outlier data sets are exist. The outlier data are where the whole bridge rating value is larger than the maximum component rating data. The missing data problem requires a method to improve the accuracy and efficiency of the modeling of the condition rating of a bridge that utilizes a mathematical model such as MRA and ANN. In terms of BMSs, inappropriate treatment of missing data affects the performance of the bridge condition rating model and the bridge deterioration predictive model.

The common methods used by previous researchers to deal with the missing data include either deleting the missing features or replacing the missing data with zero or with the mean data of the training set [5, 6]. However, as there is a certain level of uncertainty associated with a particular case, such as the bridge condition rating data, the above method must be

Rating scale	General definition
1	No damage found and no maintenance required as a result of inspection
2	Damage detected, and it is necessary to record the condition for observation purposes
3	Damage detected is slightly critical, and thus it is necessary to implement routine maintenance work
4	Damage detected is critical, and thus it is necessary to implement repair work or to carry out detailed inspection to determine whether any rehabilitation works are required or not
5	Being heavily and critically damaged, and possibly affecting the safety or traffic, it is necessary to implement emergency temporary repair work immediately, rehabilitation work without delay after the provision of a load limitation traffic sign, or replacement work

Table 1. The condition rating system based on severity of defect.

quantified in certain ways during the data-mining process. Moreover, the simple methods mentioned earlier are often not suitable for improvement of the quality of the data.

The performance of the mathematical model of bridge condition rating is also dependent on the quantity and the quality of data set used in constructing relationship between bridge component and whole bridge condition rating. Thus, the purpose of this study is to develop bridge condition rating model based on limited available data and different rates of missing bridge component condition rating data. MRA and ANN are utilized to map relationship between bridge component ratings and whole bridge condition rating. Four conditions of data sets are used: complete data sets (M0), data sets with one component missing (M1), data sets with one or two components missing (M2), and data sets with one, two, or three components missing (M3). Five methods are then used to handle the missing bridge component condition rating data. This study explores three commonly used methods and two new ones. The best method is then applied to substitute the missing data of bridge component condition ratings so that the bridge condition rating model can be developed from various conditions of data sets. Furthermore, the handling of the missing data also increases the size of data sets and provides a more complete range for the bridge condition rating distribution.

2. Previous works

Condition rating data have the potential to provide tremendous value to the bridge management. The condition rating data can be used to help prioritize maintenance work and decide on allocation of available budgets based on engineering and financial considerations [7]. Hence, the appropriate procedure is needed in bridge condition ratings data gathering for managing bridges under constrained resources.

ANNs are widely used as an attractive alternative to handle complex and non-linear systems that are difficult to model using conventional modeling techniques such as MRA. ANNs have been widely applied in engineering, science, medicine, economics, and environmental applications. The most common applications are function approximation, pattern classification, clustering, and forecasting [1, 8–10]. Various forms of ANNs (i.e., feed-forward neural networks: FENN, recurrent networks, radial basis functions, wavelet neural networks, Hopfield networks, etc.) have been applied in various disciplines. However, in the context of function approximations, such as bridge condition rating, the FFNN is generally chosen as the network architecture [9] and back propagation (BP) as the learning algorithm [10–12].

Chen [10] categorized and evaluated a beam bridge condition into four main bridge components, namely, substructure, bearing, beam, and accessory structure. These four components are evaluated based on 20 assessment criteria that can be inspected by close visual inspections according to Chinese Bridge Maintain Codes. Five neural network models are then developed to model substructure, bearing, beam, accessory structure, and whole bridge status. The input parameters for substructure, bearing, beam, accessory structure, and whole bridge status are 7, 2, 6, 5 and 4, respectively. He concluded that the proposed approach improves the efficiency of bridge state assessment.

Li et al. [13] utilized ANN to evaluate a bridge conditions based on substructure, superstructure, deck, and channel conditions. In their proposed model, the training cases converged very well, but for the test cases, the prediction from the network is consistent with the target in about 60%. They concluded that the low prediction accuracy affected by the data used in training the network is not sufficient for the network to generate proper weights to precisely model the input–output relationship. Another reason was inconsistency in the evaluation results due to subjective factors observed in the inspection data, which are used to train and test the neural network.

In most countries, there exists a large time gap between the dates of the construction of the bridge and the adoption and implementation of the relevant BMS [14]. There is a general leaking in such BMSs' database such as inconsistency of data sets and much bridge component condition rating is unrecorded. Another example of how bridge condition rating data can go missing is the difficulty in obtaining information and expensive testing of some bridge components. The missing data constitute the largest fraction of the difficulties in analyzing the data, making constructing predictive ratings, and other decision-making processes that depend on these data. Furthermore, it is impossible to build a convictive classification model with missing data because the missing data affect the integrity of the dataset [6]. Zhimin et al. [6] used five methods to handle missing data in their classification problem. The methods are as follows: deleting missing data, replacing missing data with zero, replacing missing data with the mean value of all the data of the training set, replacing missing data with the mean value from the same label data of the training set and predicting the missing value using a feed-forward backpropagation ANN. Markey et al. [5] compared three methods for estimating missing data in the evaluation of ANN models for their approximation problems. These methods are as follows: simply replacing the missing data with zero or the mean value from the training set or using a multiple imputation procedure to handle the missing value.

3. Methodology

In this study, the bridge condition ratings data were provided by the Bridge Unit, Roads Branch, PWD Malaysia. The ratings based on the Annual Bridge Inspection Manual [4] by the PWD classified the state of a bridge and bridge component conditions into five numerical systems on a rating scale of 1–5, from "no damage" (rating 1) to "heavily and critically damaged" (rating 5). The bridge components are classified as either primary or secondary elements. The primary components include the surface, deck slab, beam/girder, piers, and abutment, whereas the secondary components include the parapets, expansion joints, bearings, slope protection, and drainpipes. The bridge condition rating can be evaluated by processing the ratings and important sets of the bridge components [15]. Suksuwan [16] evaluated an overall bridge condition rating based on the condition of the superstructure and substructure. The superstructure consists of two components, namely the bridge deck and accessories. Meanwhile, the substructure is divided into three components: pier, abutment, and foundation. The relationship between the bridge condition rating and its components can be drawn as $Y = f\{X_1, X_2, \ldots, X_p\}$, where $X_1, X_2, \ldots, X_p$ are p bridge component condition rating variables and Y represents the bridge condition rating.

http://dx.doi.org/10.5772/intechopen.71556

Since the availability of data sets with complete condition data of bridge component were very limited, utilizing incomplete sets by handling missing value with the appropriate treatment is expected to improve the performance of the model. Furthermore, upon observation of the available data sets, it was found that there is a big gap in the bridge condition rating distribution, where there were no complete data sets available for a bridge condition with a rating of 4 and only four pairs of complete sets of bridge conditions with a rating of 5 available. Consequently, finding the best method to handle this problem is an important step prior to constructing the model of the bridge condition ratings.

3.1. Data preparation

The original data sets include 1244 data sets from the last 4 years of inspection records of 311 single-span concrete bridges. Among these data sets, only 579 sets have complete rating data for components and the whole bridge condition. From these 579 data sets, a large number of the bridge condition rating data sets have repeated data. When constructing the model of bridge condition ratings through an ANN and MRA, they provide nothing new as the information is redundant. To avoid this redundancy, only one of the data sets with the same data is retained and the others are deleted. Furthermore, these available data sets are also adjusted to remove the outlier data sets where the bridge condition rating value cannot be larger than the maximum component rating data. After deleting the redundant and outlier data sets, there are 157 data sets left with almost all of them having a rating of 1, 2, or 3. However, bridges with a condition rating of 4 do not have complete component rating data, and only three pairs with a rating of 5 are available, as shown in **Figure 1**.

As explained in the previous section, all data sets for bridges with a rating of 4 are incomplete. In this study, the data sets that have 1–3 components with missing value are considered and handled with different methods to provide more data and fill the gap in the bridge rating scale distribution. The number of data sets for M0, M1, M2, and M3 are 157, 226, 252, and 267, respectively.

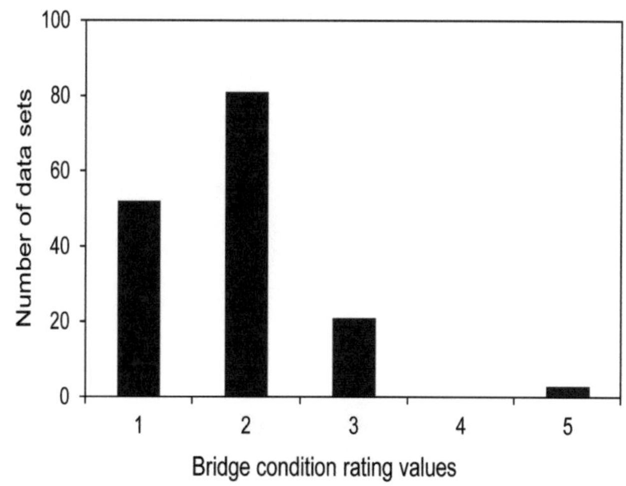

Figure 1. Bridge rating distribution of completed data sets (M0).

No. of data	Bridge component (X)					Bridge rating (Y)
	X_1	X_2	X_3	...	X_p	
1	x_{11}	x_{12}	x_{13}	...	x_{1p}	Y_1
2	x_{21}	x_{22}	x_{mis}	...	x_{2p}	Y_2
3	x_{31}	x_{32}	x_{33}	...	x_{mis}	Y_3
...	...	...	...	...	...	...
N	x_{n1}	x_{n2}	x_{n3}	...	x_{np}	Y_n

Table 2. The format of data distribution used in this study.

Table 2 illustrates data sets that contain complete data and missing data of bridge component condition ratings. Five methods are proposed to handle these missing data of bridge component ratings. These five methods are as follows: substituting with the local mean (SM), substituting with the local minimum (SMN), substituting with the local mode (SMD), substituting with the local mean value of the same component class (SMC), and substituting with the available bridge condition rating value (SBR) from the same label of data set.

In SM method, the mean value of the data label n (xn) is calculated using Eq. (1). The missing value is then substituted with the xn value. Furthermore, the local minimum value (xmin) is defined as the smallest value appearing in the same label of data set. Meanwhile, the local mode value (xmode) is defined as the value that appears most often in the same label of data set.

$$X_n = \frac{\sum_{n=1}^{p} X_{n1}}{p} \tag{1}$$

Meanwhile, in the substitution with SMC method, the missing value is substituted based on the criteria of the bridge components. If the primary component rating data are missing, the data will be substituted by the average value of other available primary component ratings. The same method is also applied to unavailable data of secondary components. By substitution with the SBR method, the missing component rating value is simply substituted with the available bridge condition rating value from the same label of data set.

After the missing data are handled with the above methods, the data are then checked to remove the redundant data that appear in the list. Furthermore, the remaining data sets show that there is no significance different in the number of data sets (M2 and M3) in comparison to M1, hence only data sets M0 and M1 are chosen in study. The distribution of data sets after removing the redundant data is shown in **Figure 2**.

3.2. Multiple regression analysis model

In this case, MRA deals with one output parameter (dependent variable), which is the bridge condition rating value and nine input parameters (independent variables) which are the bridge components condition rating. If the bridge condition rating is y and bridge components condition rating is x_1, x_2, x_3, …, x_n, then the model is given by:

$$y = \beta_0 + \beta_1 x_1 + \beta_2 x_2 + \beta_3 x_3 + \ldots + \beta_n x_n + \varepsilon \quad (2)$$

Since the data sets used in this modeling problem are more than one, then Eq. (2) can be written as follow:

$$y_i = \beta_0 + \beta_1 x_{1i} + \beta_2 x_{2i} + \beta_3 x_{3i} + \ldots + \beta_n x + \varepsilon. \quad (3)$$

where β is the coefficient of bridge components condition rating, n the number of bridge component considered, $i = 1, 2, 3, \ldots, N$ is the number of data sets, and ε represents the error of the model.

3.3. Artificial neural network model

Five conditions of data sets are then trained with a choice of the network architecture, namely complete data sets (M0), which number 157 data sets in total; data sets with one component missing (M1), which number 226 data sets in total; data sets with one or two components missing (M2), which number 252 data sets in total; and data sets with one, two, or three components missing (M3), which number 267 data sets in total.

A feed-forward neural network with a single hidden layer that varies the number of hidden neurons in the range from 1 to 28 neurons and one output layer, as shown in **Figure 3**, was selected to train the data sets. The networks are trained with a variation of the back-propagation training algorithm, namely the Levenberg-Marquardt algorithm (trainlm). The Trainlm algorithm is utilized as it has the fastest convergence in function approximation problems such as the bridge condition rating problem [17]. Bridge component condition rating data are used as input variables, which consist of nine inputs (surface, expansion joint, parapet, drainage, slopes protection, abutment, bearing, deck/slab, and beam/girder), and the whole bridge condition rating is used as the output variable Y.

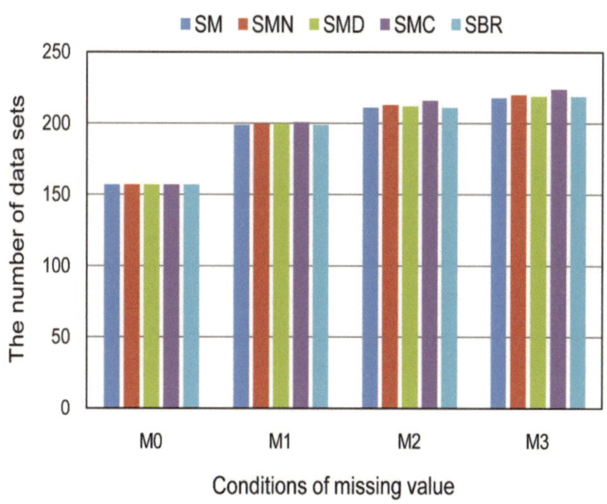

Figure 2. The distribution of data sets for data sets M0, M1, M2, and M3.

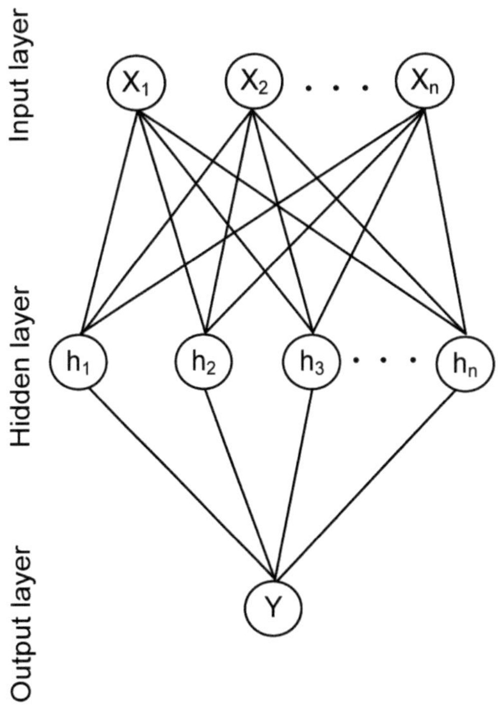

Figure 3. Structure of typical ANN model.

During the training process, there is a possible risk of overfitting or overtraining the network. In this situation, the error on the training set is driven to a very small value, but when new data are presented to the network, the error becomes large. The network has memorized the training examples, but it has not learned to generalize to new situations [17, 18]. Therefore, in this work, the early stopping technique was used to monitor the training process to handle the over-training problem. In the early stopping technique, there is a need to divide the data set into three subsets: training, validation, and testing data sets. The training set is used to train the network, and the validation data set is required to validate the network according to the early stopping technique. The testing data sets are used to test the performance of the trained network. In this study, 60, 20, and 20% were used as the training, validation, and testing data sets, respectively.

Prior to training with the data sets, the network inputs and targets are normalized using the functions [pn, ps] = mapstd(p) and [tn, ts] = mapstd(t) in MATLAB software so that they had mean of 0 and standard deviation of 1. The original network inputs and targets are given in the matrices p and t. The normalized inputs target pn and tn that are returned to have mean of 0 and standard deviation of 1. The settings structures ps and ts contain the means and standard deviations, respectively, of the original inputs and original targets. After the network has been trained, these settings are then used to transform any future inputs that are applied to the network. To convert these outputs back into the same units that were used for the original targets requires ts. The following functions simulate the network that was trained on the previous functions and then converts the network output back into the original units an = sim(net, pn) and a = mapstd("reverse," an, ts) [19].

A transfer function is used to produce the neuron output and limit the amplitude of the output of the neuron. It determines the relationship between the inputs and outputs of a neuron and a network [17]. In this study, the tangent sigmoid transfer function (tansig) and linear transfer function (purelin) are used in the hidden and output layer, respectively. The tansig function, as given in Eq. (4), produces outputs in the range of −1 to +1, and the purelin function, as given in Eq. (5), produces outputs in the range of −∞ to +∞.

$$tansig(x) = \frac{2}{(1 + exp(-2 * x))} - 1 \tag{4}$$

$$purelin(x) = x \tag{5}$$

The root mean squared error (RMSE) of the training set is used to measure the performance of the network, where the typical performance function that is used for training ANNs is the mean sum of the squares of the network errors. The coefficient of determination (R^2) of the linear regression line between the ANN outputs and the bridge condition rating targets is also used to measure the response of the trained network. The error of data label k (e_k) and RMSE of all the training sets are calculated using Eqs. (6) and (7).

$$e_k = t_k - a_k \tag{6}$$

$$RMSE = \sqrt{\frac{\sum_{i=1}^{N} (y_{target} - y_{predicted})^2}{N}} \tag{7}$$

Here, t_k is the target of data label k, a_k is the network output of data label k, and Q is the number of data sets used in the network training process.

The number of epochs for all the training algorithms is fixed at 1000. At the same time, the initial values of the weights and biases are always initiated from random values; therefore, each run might produce different output values [20]. Therefore, each ANN is made to run 30 times, and the average values for RMSE and R^2 are reported.

The process of training using the incomplete data is virtually the same as that for the training data with none of the features missing. Once the substitution has been made, the remaining steps of the training algorithm ensure the use of the handled features in the process of updating the weight and bias of the network, while the missing features are automatically handled based on the proposed solution. The purpose of the training process is to map the relationship between the input and output parameters using the ANN, as given in Eq. (8).

$$y_{predicted} = t_2[v \,.\, t_1(w \,.\, x + b_1) + b_2] \tag{8}$$

Here, x is the input vector, y is the output vector, w is the weight matrix for the connections between the input and hidden layer, v is the weight matrix for the connections between the hidden and output layer, b_1 is the bias in the hidden layer, b_2 is the bias in the output layer, t_1 is the transfer function for the neurons in the hidden layer, and t_2 is the transfer function for the neurons in the output layer [21].

4. Results and discussion

4.1. Models MRA

Table 3 presents the results of MRA models for data set with complete bridge components condition rating data and data set after substituting missing data with the proposed methods. The results show that the performance of a network trained with the entire range scale of data sets (M1) seems better than the network trained with missing range scale ratings as in data sets M0, as indicated by the R^2 values. For data sets M1, the handling method with SMD, SMC, and SBR yields almost similar R^2 values that are better in comparison to SM and SMN methods. Furthermore, in term of RMSE value, SBR method shows better performance in comparison to other methods. Referring to RMSE and R^2 value of training, validation, and training data sets, the performance of SBR method appears to yield slightly better results than other methods.

The typical MRA models linking bridge condition rating to its explanatory components for data sets M1-SBR are presented in **Table 4**. The significance of each coefficient was determined by t-value and P-value. The F-ratio presented in **Table 3** measures the probability of chance departure from a straight line. On review of the output found in **Table 3** shows that the overall model was found to be statistically significant as (F = 53.6379, Sig.-F < .0000), (F = 138.6713, Sig.-F < .0000), (F = 99.1484, Sig.-F < .0000), (F = 101.3906, Sig.-F < .0000), (F = 104.8520, Sig.-F < .0000), and (F = 148.6167, Sig.-F < .0000) for data sets M0, M1-SM, M1-SMN, M1-SMD, M1-SMC, and M1-SBR, respectively.

4.2. Models ANN

The training process is evaluated based on plotting the training, validation, and Mean Squared Errors (MSEs) versus number of epochs. **Figure 4** shows the training progress of the selected ANN for the data sets M0. The network seems unstable, as indicated in **Figure 4**, where the characteristics of the validation and test error are not similar, which may indicate that there is a gap in the data set distribution or a poor division of the data set [17].

Data sets	RMSE	F	Significance-F	R^2-Train	R^2-Val	R^2-Test
M0	0.2740	53.6379	4.31E−31	0.8518	0.8060	0.8700
M1-SM	0.2741	138.6713	1.83E−55	0.9197	0.8770	0.9140
M1-SMN	0.2816	99.1484	1.36E−48	0.9158	0.8610	0.9160
M1-SMD	0.2723	101.3906	4.58E−49	0.9213	0.8920	0.9140
M1-SMC	0.2692	104.8520	4.64E−50	0.9272	0.8830	0.8900
M1-SBR	0.2655	148.6167	5.72E−57	0.9246	0.8840	0.9140

Table 3. The statistic summaries of MRA result.

Factors	Coefficients	t-value	P-value
Intercept	−0.0658	−0.8026	0.4239
Surfacing	0.0682	2.1352	0.0350*
Expansion joint	0.0402	1.2845	0.2017
Parapet	0.0980	3.2732	0.0014*
Drainage	0.0536	1.8002	0.0746
Slope protection	0.0542	2.0419	0.0436*
Abutment	0.2138	6.1275	0.0000*
Bearing	0.0100	0.2033	0.8392
Deck slab	0.2422	6.2569	0.0000*
Beam/Girder	0.2754	6.1422	0.0000*

*Significant at $p < 0.05$.

Table 4. Model coefficients estimated by MRA for data sets M1 substituting with SBR (M1-SBR).

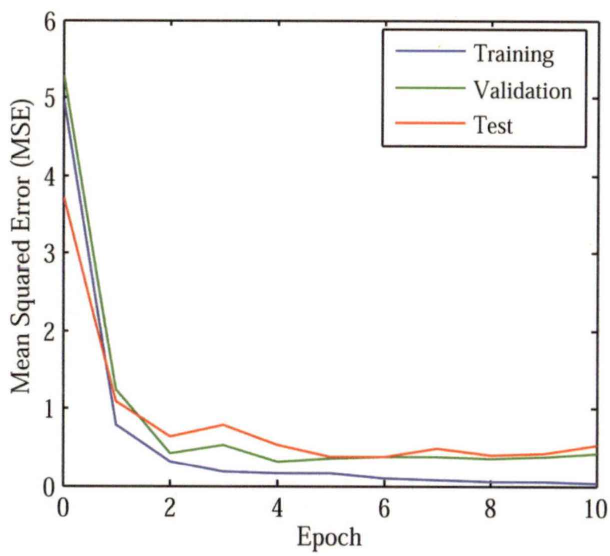

Figure 4. Typical training performance versus number epochs for data sets M0.

Figure 5 shows the typical training progress of data sets where the missing value is substituted with the bridge condition rating value (SBR). **Figure 5** shows the decrease of the MSE versus the number of epochs during the training process of the ANN. Indeed, **Figure 5** shows that the validation and testing set errors show similar characteristics, which provide reasonable evidence of network training, and it does not appear that any significant overfitting has occurred. If the error in the test set reaches a minimum at a significantly different epoch number than the validation set error, it may indicate a poor division of the data set [17].

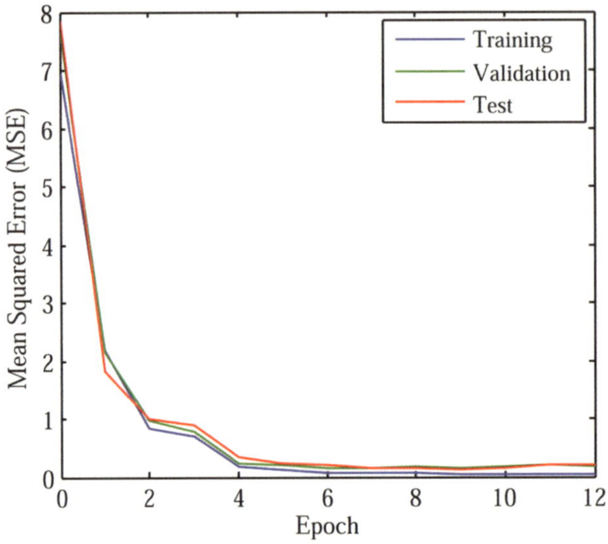

Figure 5. Typical training performance versus number epochs for data sets M1-SBR.

Regression analysis between the bridge condition rating predicted by the ANN model and the corresponding bridge condition rating target is performed using the routine postreg using MATLAB software. The format of this routine is [m,b,r] = postreg (a,t), where m and b correspond to the slope and the intercept, respectively, of the best linear regression that relates the targets to the ANN outputs. If the fit is perfect, the ANN outputs are exactly equal to the bridge condition rating targets, and the slope is 1 and the intercept with the Y-axis is 0. The third variable, r, is the correlation coefficient between the ANN outputs and targets. It is a measure of how well the variation in the predicted bridge condition rating is explained by the target. If r is equal to 1, then there is perfect correlation between the targets and ANN outputs [22].

The performance of ANN models for complete data sets (M0) and data sets (M1) after the missing data are handled by the above-mentioned methods are presented in **Table 5**. The regression analysis for data sets with the missing data substituted with the SM method yields R^2 values of 0.9375, 0.8789, and 0.9045 for the training, validation, and testing sets, respectively. The training of data sets where the missing data are substituted with the SMN method yields R^2 values of 0.9363, 0.8688, and 0.9059 for the training, validation, and testing sets, respectively. The training of data sets where the missing data are substituted with SMD method yields R^2 values of 0.9429, 0.8884, and 0.9063 for the training, validation, and testing sets, respectively. Meanwhile, the training of data sets where the missing data are substituted with the SMC method yields R^2 values of 0.9372, 0.8849, and 0.8792 for the training, validation, and testing sets, respectively. The training of data sets where the missing data are substituted with the SBR method yields R^2 values of 0.9553, 0.8922, and 0.9057 for the training, validation, and testing sets, respectively.

The results also show that the predictions of a network trained with the data sets M0 are less accurate than those of the handled data sets, as indicated by the R^2 values in **Table 5**. The R^2 values for the data sets M0 are 0.9131, 0.7515, and 0.8115 for the training, validation, and testing sets, respectively. The treated data sets yield the highest R^2 values of 0.9553, 0.8922, and

0.9057 for the training, validation, and testing sets, respectively, with the substitution of the missing value with the bridge condition rating value. The network of the incomplete data sets does not seem to perform well in fitting the entire range of the rating scale. This problem may be due to the effect of a gap between the available bridge condition rating scale (where rating 4 is unavailable) and the bridge component ratings, which have a rating scale from 1 to 5.

In terms of RMSE values, the missing bridge component data handled using SBR method yield lowest RMSE value in comparison to other substitution methods, which indicates that the SBR method improves the performance of the ANN model. The RMSE values of the data sets where the missing data are handled by SM, SMN, SMD, SMC, and SBR are 0.2349, 0.2386, 0.2257, 0.2404, and 0.2066, respectively.

The linear regression analysis of all data sets is then also performed to evaluate the developed models' response on all the data sets. Here, all the data sets, namely training, validation, and test sets, are introduced through the models, and a linear regression between the model outputs and the corresponding targets is performed. **Table 6** shows the R^2 values of all data sets for all handling methods.

The result shows that the MRA model and ANN model trained with the missing data substituted by the bridge condition rating value (SBR) has a higher accuracy of prediction in comparison to other methods, as shown in **Table 6**. The SBR method yields R^2 values of 0.9100 and 0.9328 for all data sets by MRA and ANN methods, respectively. This indicates that, for the bridge condition rating model with these data conditions, the SBR method is a more

Data sets	RMSE	R^2-Train	R^2-Val	R^2-Test
M0	0.2079	0.9131	0.7515	0.8115
M1-SM	0.2349	0.9375	0.8789	0.9045
M1-SMN	0.2386	0.9363	0.8688	0.9059
M1-SMD	0.2257	0.9429	0.8884	0.9063
M1-SMC	0.2404	0.9372	0.8849	0.8792
M1-SBR	0.2066	0.9553	0.8922	0.9057

Table 5. The ANN models performance for data sets M0 and M1.

Conditions of data sets	MRA models	ANN models
M0	0.8380	0.8605
M1-SM	0.9050	0.9192
M1-SMN	0.8980	0.9167
M1-SMD	0.9090	0.9247
M1-SMC	0.9040	0.9151
M1-SBR	0.9100	0.9328

Table 6. The R^2 value of entire data sets for all substituting methods.

reasonable method for handling the missing value prior to training the network. The performance comparison of ANN models and MRA models is also made in terms of R^2 values of predicted value versus target value as shown in **Table 6**. It can be seen that all ANN models provide better agreement with the target of bridge condition rating.

5. Conclusions

Bridge condition rating models are developed based on very limited inspection data records using MRA and ANN techniques. Since the data sets are limited, utilizing all the available data sets by handling missing data with the appropriate methods is expected to improve the performance of the models. The method, where the missing value is substituted with the bridge condition rating value, performs better than the other methods. This method is able to determine the bridge condition rating with R^2 values of 0.9553, 0.8922, and 0.9057 for the training, validation, and testing data sets, respectively, by ANN technique. It can be concluded that constructing a model with a complete range of the rating scale is more reasonable for bridge condition rating problems compared with constructing the model using only the available rating scale. Furthermore, there was no significant difference between R^2 value of validation and testing set for treated data sets in comparison to data sets M0. The R^2 values for validation and testing set of data sets after missing data that are substituted by SBR are 0.8922 and 0.9057, respectively. Meanwhile, the R^2 values for validation and testing sets of data sets M0 are 0.7515 and 0.8115, respectively. It can also be concluded that the ANN models perform slightly better than MRA in mapping relationship between bridge components and bridge condition rating.

Acknowledgements

The authors wish to acknowledge the financial support from the Ministry of Science, Technology and Innovation (MOSTI) of Malaysia for this work, which forms part of a project on the "Development of condition rating procedure of integral bridges" (Project Number: 06-01-02-SF0323) and Ministry of Higher Learning Malaysia for grant FRGS/1/2016/TK06/UKM/02/2. The authors also thank the Public Works Department of Malaysia for providing the data.

Author details

Khairullah Yusuf[1,2] and Roszilah Hamid[2*]

*Address all correspondence to: roszilah@ukm.edu.my

1 Department of Civil Engineering, Faculty of Engineering, Malikussaleh University, Indonesia

2 Department of Civil and Structural Engineering, Universiti Kebangsaanm Malaysia, Malaysia

References

[1] Lee J. A Methodology for Developing Bridge Condition Rating Models Based on Limited Inspection Records [Dissertation]. Australia: Griffith University; 2007

[2] Rashidi M, Gibson P. Proposal of a methodology for bridge condition assessment. In: Australasian Transport Research Forum 2011 Proceedings; Adelaide, Australia. 2011. p. 1-13

[3] Saito M, Sinha KCA. Delphi study on bridge condition rating and effects of improvements. Journal of Transportation Engineering. 1991;**117**(3):320-334

[4] PWD. Annual Bridge Inspection Manual. Malaysia: Bridge Unit, Roads Branch, Public Work Department; 2003

[5] Markey MK, Tourassi GD, Margolis M, DeLong DM. Impact of missing data in evaluating artificial neural networks trained on complete data. Computers in Biology and Medicine. 2006;**36**(5):516-525

[6] Zhimin M, Zhisong P, Guyu H, Luwen Z. Treating missing data processing based on neural network and AdaBoost. In: IEEE International Conference on Grey Systems and Intelligent Services; IEEE; 2008. p. 1107-1111

[7] Saleh AM, Mansour OA, Abbas OS. Toward an Egyptian bridge management system. Housing and Building National Research Center Journal. 2013;**9**(3):227-234

[8] Jain AK, Mao J, Mohiuddin K. Artificial neural networks: A tutorial. IEEE Computer. 1996;**29**(3):31-44

[9] Shamseldin A, Nasr A, O'Connor K. Comparison of different forms of the multi-layer feed-forward neural network method used for river flow forecasting. Hydrology and Earth System Sciences. 2002;**6**(4):671-684

[10] Chen M.A. Neural network approach for existing bridge evaluation based on grid. In: Second International Symposium on Intelligent Information Technology Application; 20-22 December; Shanghai, China. IEEE; 2008. p. 90-93

[11] Kawamura K, Miyamoto A. Condition state evaluation of existing reinforced concrete bridges using neuro-fuzzy hybrid system. Computers & Structures. 2003;**81**(18):1931-1940

[12] Wang YM, Elhag TMS. A comparison of neural network, evidential reasoning and multiple regression analysis in modelling bridge risks. Expert Systems with Applications. 2007;**32**(2):336-348

[13] Li Z, Shi Z, Ososanya ET. Evaluation of bridge conditions using artificial neural networks. In: Proc. IEEE; Piscataway, NJ; 1996. p. 366-369

[14] Lee J, Sanmugarasa K, Blumenstein M, Loo YC. Improving the reliability of a bridge management system (BMS) using an ANN-based backward prediction model (BPM). Automation in Construction. 2008;**17**(6):758-772

[15] Sasmal S, Ramanjaneyulu K. Condition evaluation of existing reinforced concrete bridges using fuzzy based analytic hierarchy approach. Expert Systems with Applications. 2008; **35**(3):1430-1443

[16] Suksuwan N, Hadikusumo BHW. Condition rating system for Thailand concrete bridges. Journal of Construction in Developing Countries. 2010;**15**(1):1-27

[17] Demuth HB, Beale M. Neural Network Toolbox: For Use with MATLAB. Citeseer; 2000

[18] Zhang Z, Friedrich K. Artificial neural networks applied to polymer composites: A review. Composites Science and Technology. 2003;**63**(14):2029-2044

[19] http://www.mathworks.es/access/helpdesk/help/toolbox/nnet/backpr10.html [Internet]. [Internet]

[20] Ahadiana S, Moradiana S, Sharifa F, Tehran MA, Mohsenia M. Application of artificial neural network (ANN) in order to predict the surface free energy of powders using the capillary rise method. Colloids and Surfaces A: Physicochemical and Engineering Aspects. 2007;**302**(1-3):280-285

[21] Bhattacharjy RK, Datta B, Satish MG. Performance of an artificial neural network model for simulating saltwater intrusion process in coastal aquifers when training with noisy data. KSCE Journal of Civil Engineering. 2009;**13**(3):205-215

[22] Hamid R, Khairullah Y, Khalim AR. Influence of the primary bridge component condition on the overall bridge condition rating. In: Caner A, Gulkan P, Mahmoud K, editors. Developments in International Bridge Engineering. Part of the Springer Tracts on Transportation and Traffic Book Series. Switzerland: Springer International Publishing; 2016. p. 177-186

Chapter 4

Structural Identification (St-Id) Concept for Performance Prediction of Long-Span Bridges

Selcuk Bas

Additional information is available at the end of the chapter

http://dx.doi.org/10.5772/intechopen.71558

Abstract

Long-span cable-supported bridges are the lifeline structures for the transportation network in a country/state. An effective solution of this type of bridges is therefore indispensable not only to better understand structural response of them but also to conduct an efficient maintenance and management strategy for these bridges. In this study, structural identification (St-Id) is implemented to estimate the performance of the Bosphorus Bridge. In addition, certain efforts from finite element modeling (FEM) to utilization for performance prediction are given based on each step of St-Id. St-Id concept is divided into two main parts: experimental and numerical investigations. Due to the high cost and time limitation for testing of long-span bridges, the most effective solution to the experimental research is SHM system (SHMs). For this purpose, the SHMs of the Bosphorus Bridge is considered, finite element modeling provides an extended solution from analysis to model updating of the bridges. Considering structural performance of the bridge under extreme wind load and multi-point earthquake motion is estimated. The results from the current study indicate that St-Id concept is a robust approach for overall structural condition assessment and performance prediction of long-span cable-supported bridges.

Keywords: structural identification (St-Id), long-span bridges, structural health monitoring (SHM), finite element model (FEM), FEM calibration, multi-point earthquake analysis (Mp-sup)

1. Introduction

Long-span bridges, such as cable-stayed and suspension bridges having considerably different structural system compared with other structures are one of the main components in the transportation network in a city/state. Therefore, the governments make sizeable investment both for their construction and maintenance costs as wells as rehabilitation and retrofit projects. Considering such crucial functions of them and various investments for improving their structural performance, it is clear that comprehensive investigations on long-span bridges are necessary before and after its construction since they fall outside the scope of general bridge codes.

Over the last decades, the engineers and researchers have focused on both experimental and analytical studies taking the complexity and non-linear characteristics of long-span cable-supported bridges into consideration. For analytical studies, these structures are modeled numerically by finite element model (FEM). With utilizing FEM method, modal analysis, buckling analysis and earthquake-induced analysis of complex structures are readily carried out. Although this sophisticated analytical modeling is a powerful tool for simulating three dimensional local and global behavior of long-span bridges, it does not reflect the real structural behavior owing to uncertainty in idealization and assumption, such as in material model and analysis method. Therefore, only FE model is not enough for fully understanding the response and behavior of this type of large-scale structures.

Experimental field studies of long-span cable-supported bridges give reliable consequences for the current performance of them. Therefore, this effort is relatively significant to compare the outcomes from FEM analysis with those obtained from experimental field studies; however, this analysis could not be possible many times, especially for long-span cable-supported bridges. Additionally, data analysis and management are necessary for interpretation of collected experimental data. For this objective, structural health monitoring (SHM) is proposed as a complementary experimental technique, including structural condition evaluation, model development and calibrating, and real-time monitoring. With the major leaps in sensing and information technology, utilizing SHM to properly assess new designed and existing structures is growing interest worldwide though a standard approach for SHM has not been available yet.

In an effort to deal with the incompatibility between measured and simulated response of long-span cable–supported bridges and make an overall condition assessment of these bridges, structural identification (St-Id) concept presents an integrated approach. Structural identification is the process of developing calibrated/updated analytical model (mathematical or geometric) of structures from experiments and observations for reliable performance evaluation, simulation and decision-making. Structural identification (St-Id) and system identification (Sys-Id) are two concepts that are closely related to each other. Therefore, St-Id is a result of translation and application of system identification to mechanical and civil structural system [1]. This concept was first applied to engineering mechanics [2] and to civil-structural engineering [3]. Although many efforts are made to implement St-Id to structural engineering field, applications of this concept have not been wide-spread in civil engineering practice yet [4]. Depending upon developments in sensing and information technology, St-Id concept becomes more popular for confidential performance assessment and simulation of structures [5]. For different aspect of the St-Id concept, a number of studies were found in literature. The key point in these studies is that structural health monitoring (SHM) is utilized for experimental step of St-Id.

The most comprehensive study for St-Id was carried out by the ASCE SEI Committee on Structural Identification of Constructed Systems [1]. In addition to detailed knowledge for this process given in this study, a set of case studies are presented to illustrate the application of St-Id to building and bridge structures. Six chapters are introduced to properly comprehend St-Id concept. The scope of these sections is related to modeling, experimentation, data processing, comparison of model and experiment and decision-making. As schematically shown in **Figure 1**, St-Id consists of six steps: (i) Visualization and conceptualization (ii) Priori modeling (iii) Experimentation (iv) Data analysis (v) Calibration (vi) Utilization.

In addition to theoretical aspects, St-Id is widely utilized from estimating structural condition evaluation, rehabilitation to SHM system improvement and optimization of scaled and existing/new constructed long-span bridges under different loading and environmental conditions. A St-Id study on a newly constructed arch bridge in China was conducted in order to make FE model updating of the bridge through certain data-driven methods [6]. They demonstrated that the updated numerical model could be reliably adopted for the detailed structural analysis of arch bridges. Similar study was made in [7]. They proposed a method of frequency response function parameter identification for full-scale bridge using field-testing data. The developed approach to the model updating/calibration of the bridge is indicated to be confidently considered for FE model calibration of full-scale bridges. For various objectives of structural identification, many studies were carried out from the numerical model calibration to SHM system modification/improvement in [8–13]. In these studies, either all or certain steps of St-Id concept are aimed to be identified especially for long-span cable-supported bridges.

The Bosphorus Bridge as shown in **Figure 2** also called the 1st Bosphorus Bridge is one of the three long-span bridges in Turkey, which provides a connection between two continents of

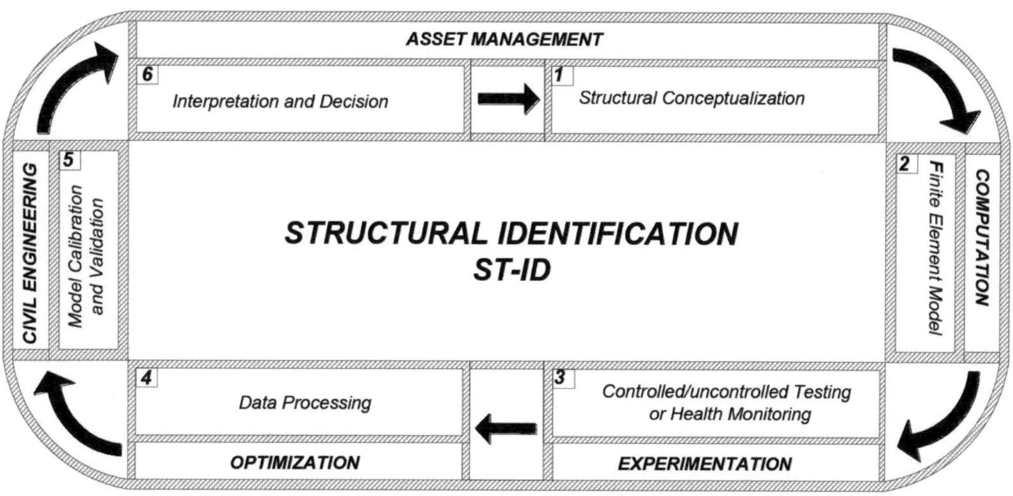

Figure 1. Layout for structural identification (St-Id).

Figure 2. General views of the Bosphorus Bridge.

the Asian and the European. When opened to traffic in 1973, the bridge was classified as the 4th longest suspension bridge in the world according to its main span. The bridge serves as vital link on the Motorway-1 (O1) connecting the city center of Istanbul. Significant part of heavy traffic of Istanbul has been carried from the bridge along with the Fatih Sultan Mehmet Bridge named the 2nd Bosphorus Bridge located on the northern side of the Bosphorus Bridge.

Taking the efficiency of St-Id into account, this approach is utilized for the Bosphorus Suspension Bridge and the results are presented in this study. Each step of the concept is handled to be conducted for the bridge. Along with establishing the numerical model of the bridge to obtain its analytical modal frequencies, the experimental study is conducted using SHM data collected during the extreme wind event. Based on the discrepancy between FEM and experimental results, model verification/calibration is performed. The calibrated model is then used for utilization of multi-point earthquake analysis of the bridge. The outcomes from this study illustrated that St-Id concept is a robust methods to properly predict the structural performance of the Bosphorus Bridge.

2. Finite element modeling and modal analysis

As in many disciplines, the application of a detailed accurate finite element model in bridge engineering field has been relatively significant to properly assess structural performance of bridges. The major leaps in computation technology enable to make advanced FE modeling and analysis. Due to the complexity of long-span cable-supported bridges, these structures are modeled based on certain assumptions: (a) simplified spine-beam model, (b) multi-scale (hybrid) FE model. The spine-beam model that enables to reduce degree-of-freedom (DOF) of bridge structures presents the results from the global behavior of bridges. Using the spine-beam model, modal analysis can be readily conducted to obtain the effective mode frequencies and associated mode shapes. In addition, preliminary FE model that is developed based on as-built project drawings can also be validated using this modeling technique. Therefore, the spine-beam modeling technique helps to check the considered equivalent sectional properties of structural components. On the other hand, multi-scale modeling becomes very significant upon making analysis for local structural component. This modeling technique also provides that different element types, beam, solid, shell, and truss, are used together to establish 3D full-scale bridge FE model. Accordingly, both modeling techniques are utilized for different goals, and generally, they have been used together for structural analysis of bridge structures.

The modal analysis of structures is a powerful tool for earthquake excitation analysis of structures. Through this analysis, the response of structures to dynamic input can be estimated and certain outcomes related to dynamic inputs can be explained [17]. For large-scale bridge structures with different large size of structural component, such as main deck, tower etc., the mode shapes may show which component dominates the dynamic response of long-span bridges [14–16].

The bridge's structural components of the tower, the main deck, the portal beams, and approach span are modeled as equivalent frame element corresponding their mechanical

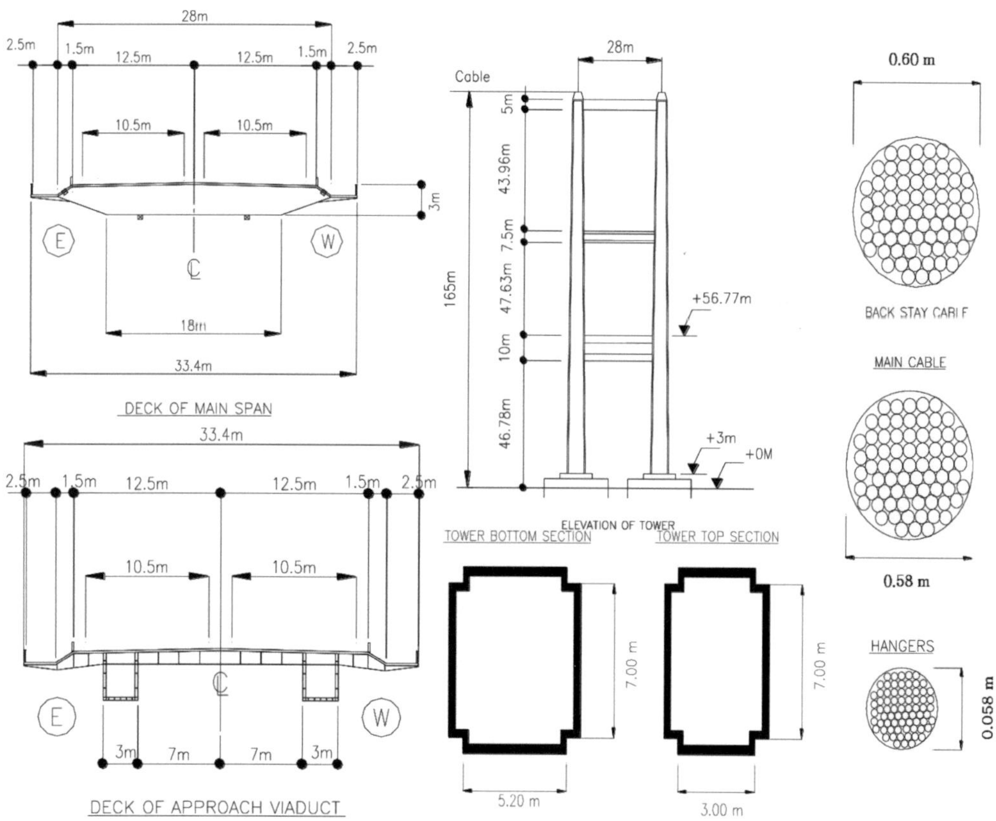

Figure 3. Sectional properties of the Bosphorus Bridge [17].

and sectional properties. The details of the structural components of the bridge are given in **Figure 3**. For elaborate sectional properties, all points of the components are precisely determined depending on the project drawings, and thus much more realistic dimensions of them are adopted. Based on these project specifications and general properties of the Bosphorus Bridge, FE model of the bridge is developed utilizing the spine-beam modeling approach as shown in **Figure 4**.

Since the bridge was made of structural steel, modal damping ratio of $\xi = 0.02$ is also considered to calculate the proportional structural damping for the bridge. The first 50 natural frequencies and associated mode shapes are obtained and the first five modes and associated frequencies are shown in **Figure 5**. From the analysis, the main deck of the bridge is obtained to be effective for lateral and vertical response of the bridge to a dynamic input. Particularly, modal participating total mass ratio for transverse direction of the main deck is determined as 60% at the end of the first five modes directly pertinent to the main deck mode shapes. Compared to modal participating total mass ratio of 96% at the end of the fifty modes, this value indicated the efficiency of the main deck mode shapes on the dynamic response of the bridge. Similar single mode shapes are also determined for the tower after the main deck mode shapes. All these single mode shapes of the main deck and the tower are seen in the first ten mode shapes. The other mode shapes are obtained as the combination of these single mode shapes. Based on these consequences, the main deck and the tower dynamic response are estimated to dominate the dynamic behavior of the Bosphorus Bridge. The obtained analytical modal frequencies are also compared with those from the experimental results so as to verify/calibrate FEM of the bridge.

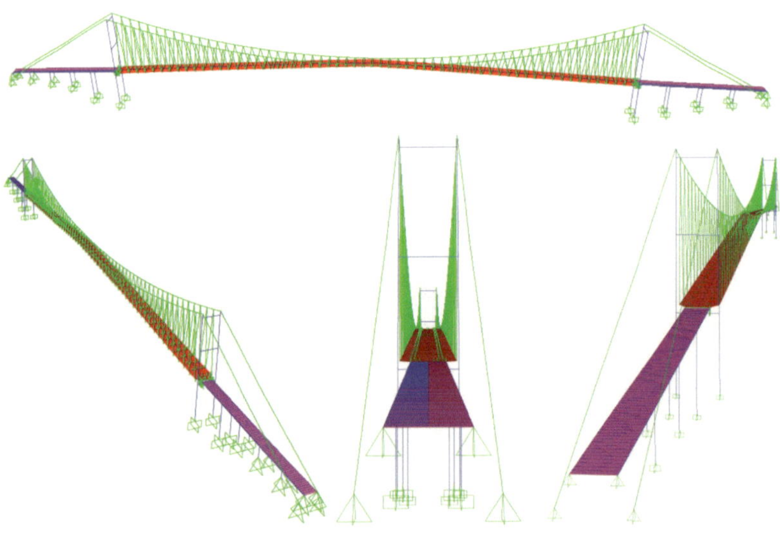

Figure 4. 3-D full-scale FE model of the Bosphorus Bridge.

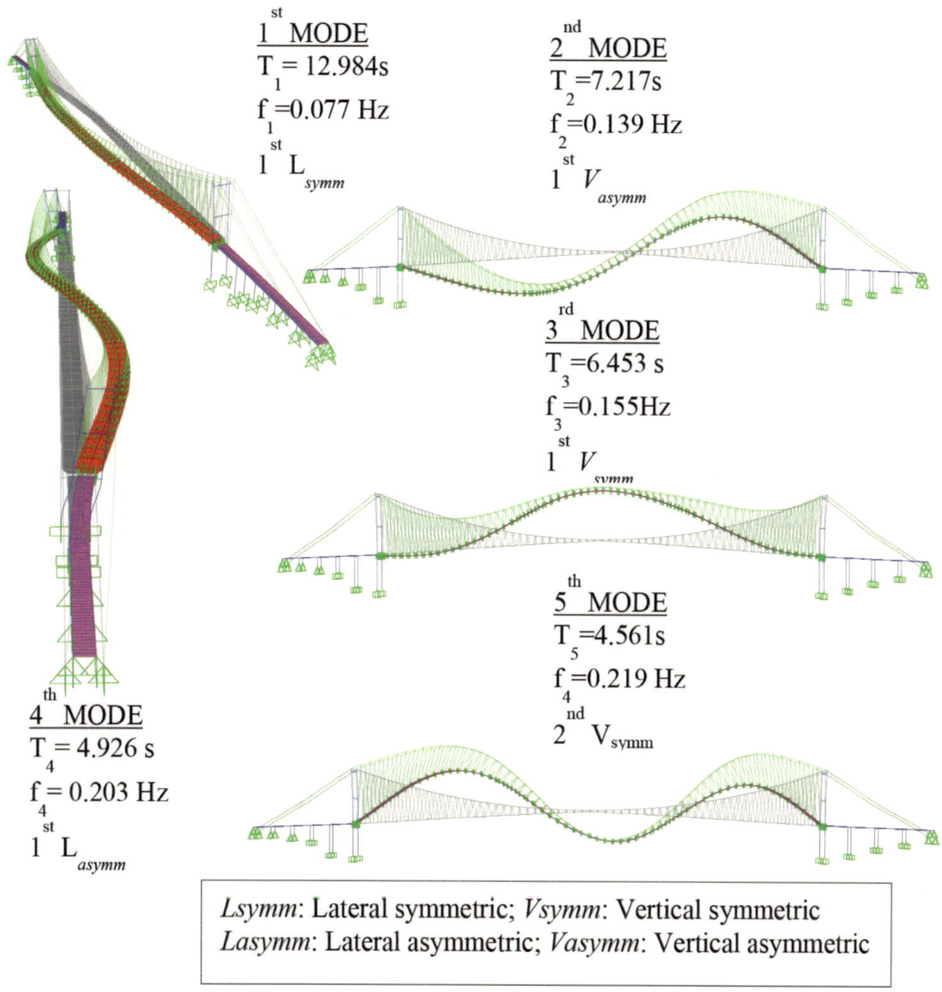

Figure 5. The first five mode frequencies and corresponding shapes of the Bosphorus Bridge.

3. Structural health monitoring system (SHM)

In general, the components of a SHM system are; (i) Sensory systems (ii) Data acquisition and transmission systems (iii) Data processing and control system (iv) Data management systems (v) Structural evaluation system [18]. Design of SHM systems is based on clearly describing the monitoring objectives. Therefore, working SHM designer together with bridge designer is inevitable to identify the objectives very well. Considering the objectives, monitoring requirements should be properly identified [19]. The requirements can be considered as below;

- The parameters to be monitored (wind, displacement, temperature)
- The design value and measurement range of the parameters
- The spatial and temporal properties of the parameters
- The accuracy requirements
- The environmental condition of the monitoring
- The duration of the monitoring

Following identification of the monitoring parameters, the next step is to determine the number of sensors based on the monitoring objectives. In order to compensate for the requirements of the monitoring, it is important to select which types of sensor are used. For this purpose, technical specifications of the sensors including measurement range, sampling rate, sensitivity, resolution, linearity, stability, accuracy, repeatability, frequency responses should be identified clearly.

With the daily traffic capacity of 195 thousands vehicles, the Bosphorus Bridge serves as its significant function without no traffic interruption in the transportation network of Istanbul. Due to the seismicity and location of Istanbul, the bridge has been subjected to many critical loading events from earthquake, wind, heavy traffic to marathon. Therefore, it is a crucial issue to track the structural response of the Bosphorus Bridge under these type of excitations. The bridge was decided to be donated with SHM system due to the unpredicted failure of hanger rope in 2004. With the total number of 258 channels and 168 sensors varying from accelerometers, tilt meters, force transducers, strain gauges, laser displacement, GPS, thermocouples to weather stations, a permanent structural health monitoring system (SHM) is developed considering the bridge's own characteristics and critical points to be monitored [20].

The quantity and location of the sensors are determined after certain temporal installation tests on its SHM system. In **Table 1**, the number and type of sensors installed on the bridge are listed with their preferences. Moreover, general sensor arrangement of the SHM system of the bridge is presented in **Figure 6**.

Type	Preferences	Quantity
Accelerometers	Measuring range of +/− 2 g	19
Tiltmeters	Measuring range (°): ±14.50	15
Force transducer	Measuring range (mm): ±1.50 mm	12
Strain gauges	Resistance tolerance (%): ±0.30	70
Laser displacements	Measuring range (mm): 200–2000	8
GPS	Precision (mm): 0.2	5
Thermocouples	Accuracy (%): ±0.10	33
Weather station	Wind speed range (mph): 0–130	6
Total		168

Table 1. Sensor types and quantity of the SHM system of the bridge [20].

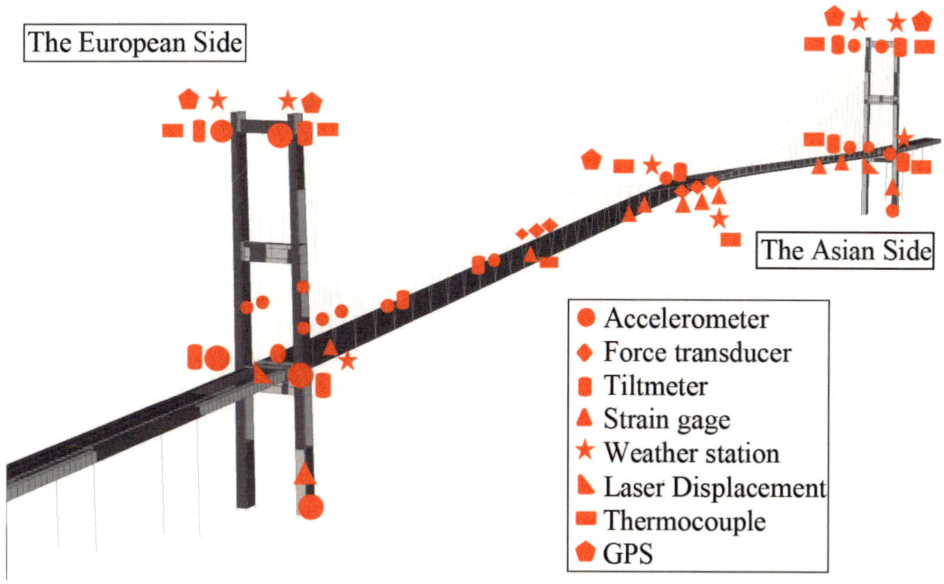

Figure 6. General layout for the sensors of the SHM system of the bridge.

4. Structural identification (St-Id)

4.1. Description of extreme wind event

Strong winds are not very frequent in Istanbul due to its location. However, during the daytime on April 18 2012, a strong storm occurred in Istanbul. It was the first time that the bridge experienced such a high wind. According to measurement of Turkish Meteorology Service, the peak wind-speed reached to 122 km/h. Although ultimate design wind speed of the bridge is 162 km/h, the bridge was closed to the traffic for a period of time as a precautionary measure. The change of wind speed with time is shown in **Figure 7**. This variation is also obtained by weather stations installed on the bridge. As seen from **Figure 7**, the mean wind speed before the storm was around 20 km/h. However, it suddenly increased to 100–120 km/h in 10 minutes [21]. This variation is also verified with meteorology data as shown in **Figure 7**. In addition, the excitation of the strong wind load is also corrected with the other weather station data recorded from different critical points of the bridge. As seen from **Figure 8**, the lateral wind direction is verified with the polar charts through SHM data at the deck mid-span, and the bridge was determined to be greatly induced in N-S direction during the critical wind event.

4.2. System identification of the bridge

Before the identification, data processing on the acceleration data is implemented. Firstly, base-line correction is performed to get rid of offset value. The next step is to remove linear trends from the data using de-trend technique. Besides, the mean value (DC) of the data

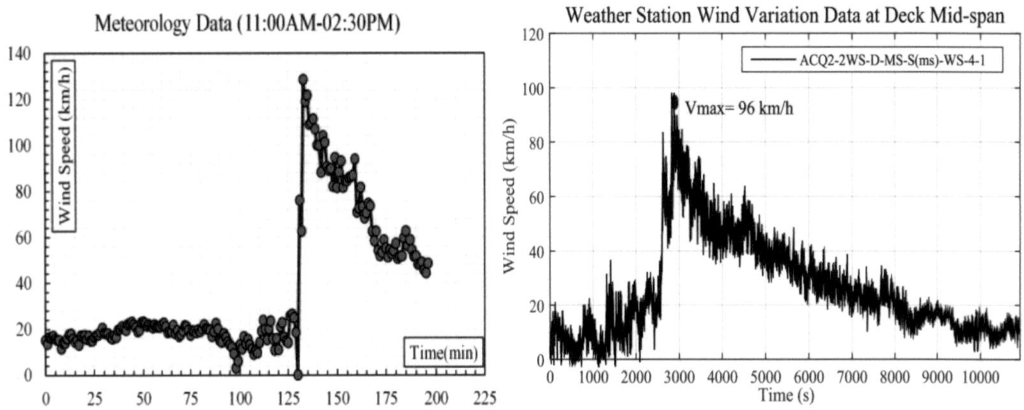

Figure 7. Identification of strong wind: meteorology data and SHM data.

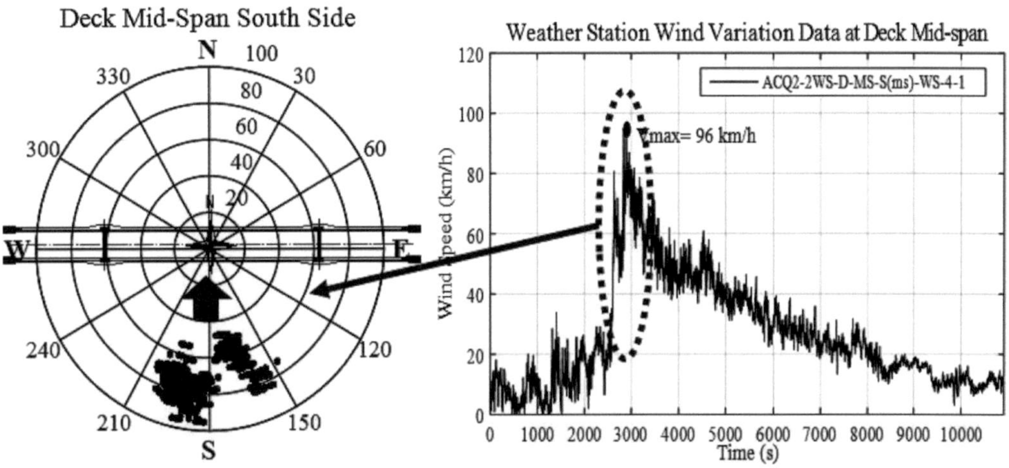

Figure 8. Identification of wind direction from the deck mid-span SHM data.

is subtracted. In order to remove the unwanted noise component, a standard fourth order Butterworth band-pass filter with the first corner frequency of 0.05 Hz and the second corner frequency of 5.0 Hz is performed [21]. For all these works, MATLAB computing program developed by the Math Works [22] is utilized. For more refined results from FFT (Fast Fourier Transform), data averaging technique including windowing and overlapping is also implemented. Window length is determined considering the minimum frequency range of the modal values obtained from the previous experimental study for the bridge in literature.

In order to determine the effects of the strong wind on dynamic characteristics and operational performance of the bridge, all SHM data are divided into three ranges as shown in **Figure 9**: "Before", "During" and "After". The data analysis shows that meaningful results are not

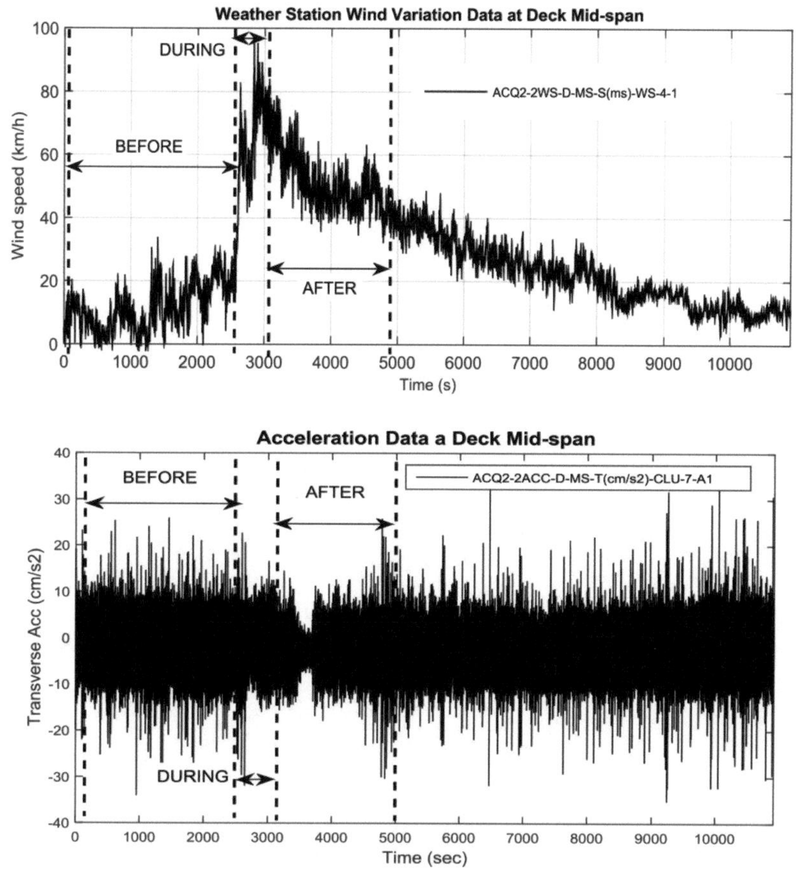

Figure 9. Three ranges on SHM data: "Before", "During" and "After".

obtained for "Before" range since the data is relatively distorted with traffic noise. Therefore, "After" range is considered for the comparison with "During" range. The obtained SHM data from the accelerometers mounted at the specific points of the bridge are utilized for structural identification of the bridge. The locations of the accelerometers are indicated with the considered directions in **Figure 10**. In this figure, the FFT analyses of the all accelerometers are also given for "During" range to determine modal characteristics of the bridge. This effort is repeated again for "After" range and the outcomes from frequency-domain analysis are summarized in **Table 2**.

As mentioned previously, the results from "After" range can be considered as natural vibration characteristics of the bridge. In addition to the other studies in literature, "After" range provides an opportunity to make calibration of the developed FE model. Therefore, the comparison of the outcomes from "After" with those from the developed FE model is made as given in **Table 3**.

According to % change in **Table 3**, the maximum error % in terms of frequency is obtained for Mode-2 with approximately 12%. For the other modes, the error is in the range of 0–3.0%.

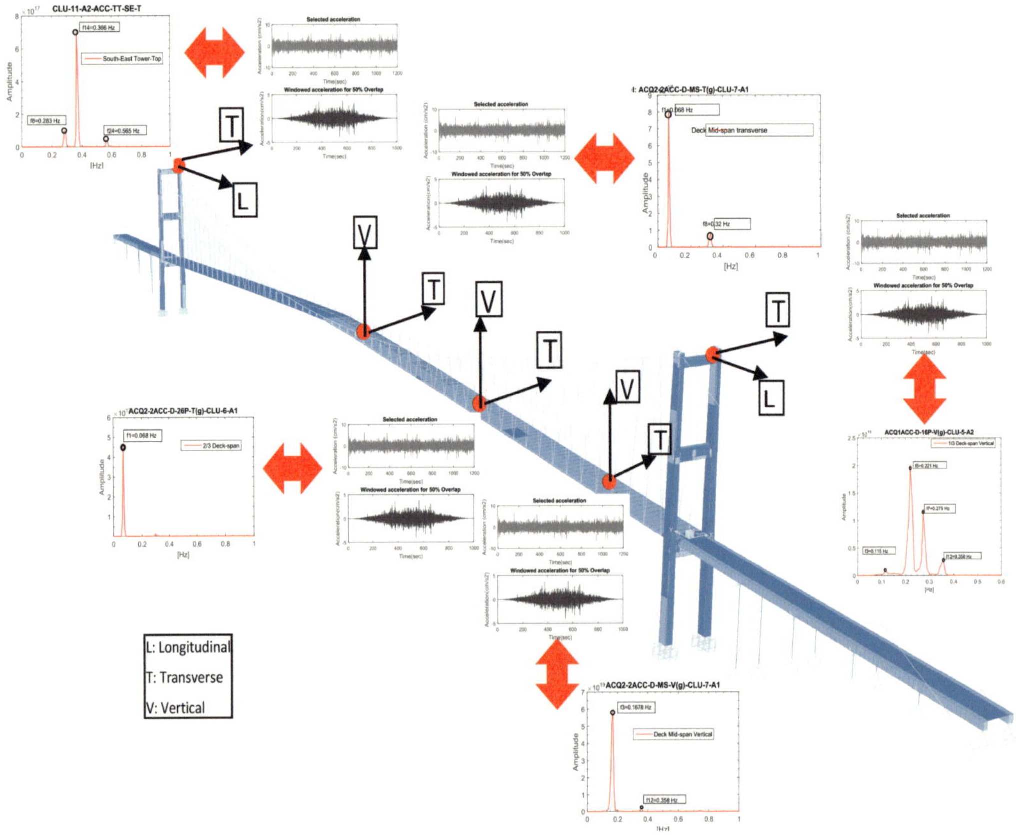

Figure 10. The FFT analysis of the extreme wind data for "During" range.

Mode-2 is identified only from "During" range instead of "After" range considered for ambient vibration. Therefore, such high error in Mode-2 is dependent on that the mode is not be identified from "After" range. Generally, the error range for the first five mode shapes of the bridge is allowable level. These conclusions reveal that the considerations utilized in FE mod-

Mode number	Mode shape	Frequency/period [Hz]/[s]					
		During		After		Change (%)	
		Period [s]	Freq. [Hz]	Period [s]	Freq. [Hz]	Period [s]	Freq. [Hz]
Mode-1	1st L_{sym}	14.706	0.068	12.766	0.078	15.196	−13.191
Mode-2	1st V_{asym}	8.065	0.124	7.853	0.127	2.698	−2.362
Mode-3	1st V_{sym}	6.096	0.164	6.250	0.160	−2.469	2.531
Mode-4	1st L_{asym}	4.854	0.206	4.967	0.201	−2.265	2.318
Mode-5	2nd V_{sym}	4.561	0.219	4.525	0.221	0.798	−0.792

L_{sym}: Lateral symmetric; L_{asym}: Lateral asymmetric; V_{sym}: Vertical symmetric; V_{asym}: Vertical asymmetric.

Table 2. FFT analysis results for "During" and "After" ranges.

Mode number	Mode shape	FE model and experimental					
		FEM		Experimental		Change (%)	
		Period [s]	Freq. [Hz]	Period [s]	Freq. [Hz]	Period [s]	Freq. [Hz]
Mode-1	1st L_{sym}	12.984	0.077	12.766	0.078	1.707	−1.679
Mode-2	1st V_{asym}	7.217	0.139	8.065	0.124	−10.505	11.738
Mode-3	1st V_{sym}	6.453	0.155	6.250	0.160	3.251	−3.149
Mode-4	1st L_{asym}	4.926	0.203	4.967	0.201	−0.819	0.825
Mode-5	2nd V_{sym}	4.561	0.219	4.525	0.221	0.809	−0.803

L_{sym}: Lateral symmetric; L_{asym}: Lateral asymmetric; V_{sym}: Vertical symmetric; V_{asym}: Vertical asymmetric; T: Torsional.

Table 3. Comparison of FEM with those from experimental results.

eling of the Bosphorus Bridge are estimated well and are properly implemented. Accordingly, the developed FE model can be reliably utilized for advanced analysis of the bridge.

5. Utilization, decision-making and performance prediction

After the destructive earthquakes in last 2 decades in Turkey, Izmit (1999) and Duzce (1999) earthquakes, the public awareness of structural earthquake safety and performance of the existing structures in Turkey has increased progressively. General Directorate of Turkish State Highways (KGM) conducted a number of rehabilitation projects (JBSI) [23] for the most critical long-span bridges in Turkey, such as the Bosphorus Bridge. The related studies for the Bosphorus in literature were basically focused on the uniform support earthquake analysis (U-sup) of the bridge. Therefore, the multi-point earthquake analysis (Mp-sup) is required to better understand the seismic behavior of the Bosphorus Bridge. Considering these recommendations, this chapter aims at determining the effects of spatially varying earthquake motion on the Bosphorus Bridge using the calibrated FEM of the bridge in the previous section.

In order to simulate site-specific ground motions, the geographic coordinates of the bridge's support points have firstly to be determined. As indicated in **Figure 11**, the support coordinates of the bridge are obtained depending on the general coordinates of the bridge. **Figure 11** also presents the general considerations of the multi-point earthquake analysis of the bridge. Taking the scenario earthquake of Mw = 7.4 predicted to occur with the probability of 70% in next 30 years in Istanbul and these coordinates of the bridge into consideration, the stochastic modeling technique proposed in [24] is used to generate spatially varying site-specific earthquake ground motions.

The simulation process is performed and the acceleration ground motion time-histories (ATH) are generated for the Bosphorus Bridge. Although the process yields to the ATHs, the displacement ground motion time histories (DTH) need to be obtained for the multi-point earthquake analysis. Therefore, the DTHs are presented in **Figure 12** instead of the ATHs. As shown in **Figure 12**, the triple-direction (two horizontals and one vertical) ground motions are generated for the each considered multi-point, A, B, C and D. Total number of twelve ground motions are defined for the analysis.

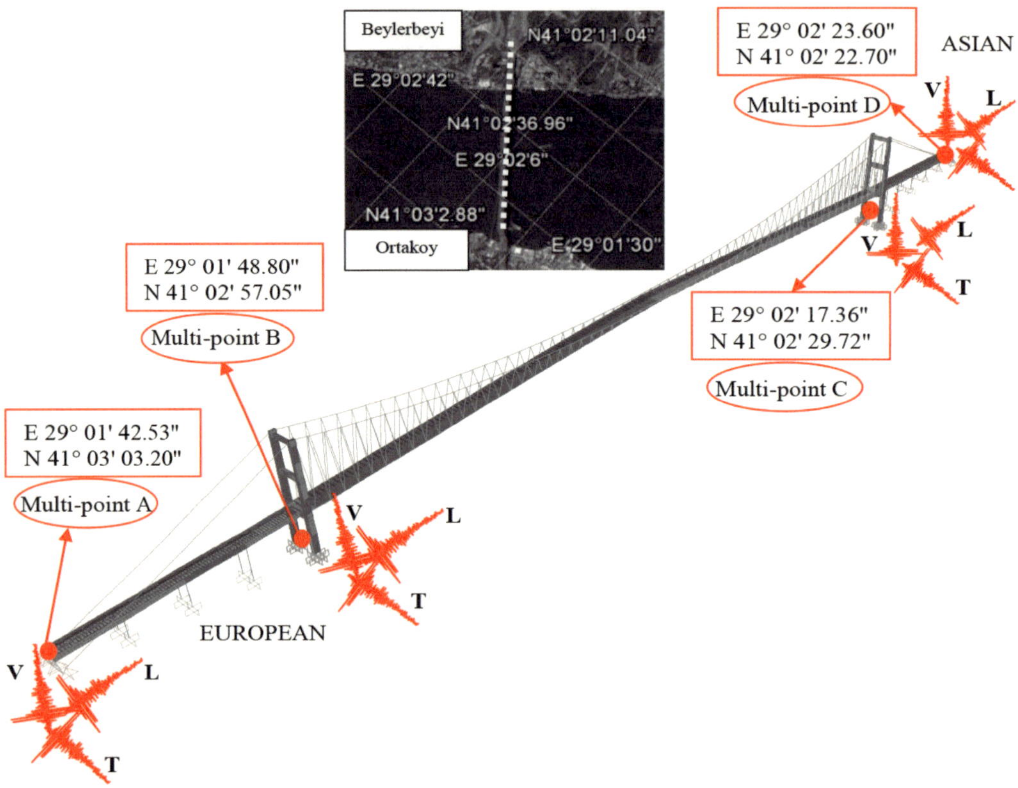

Figure 11. Geographic coordinates and multi-support earthquake analysis considerations.

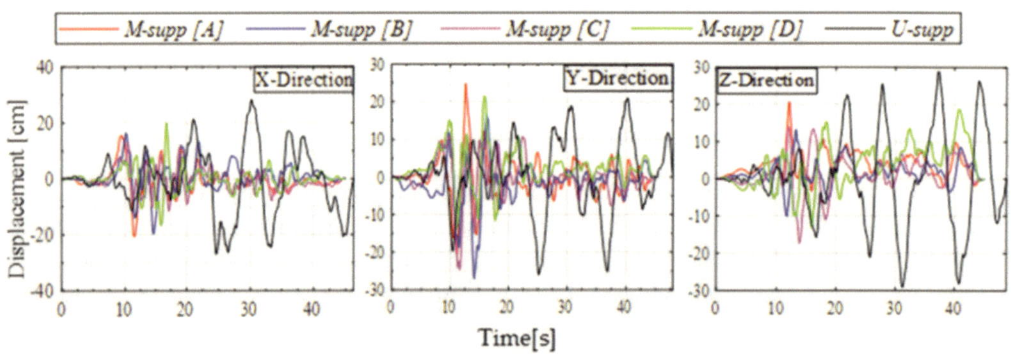

Figure 12. Simulated spatially varying site-specific multi-point earthquake records.

In **Figure 13**, the variation of the tensile strength of the main and the back stay cables is presented. The tensile strength value of the main cable increases as 74% and 78% under the Mp-sup compared to the U-sup and JBSI retrofit project, respectively. The different percentage change reveals that JBSI retrofit project is highly conservative in terms of sectional forces. Similar percentage change is obtained for the back-stay cable as shown in

Figure 13. All these results demonstrate the importance of the behavior of the deck. As to the main cable, the axial force of the main cable at the tower top-saddle also increases relatively. This increase is mostly related to the displacement of the deck and high increase in the tensile strength of the main and the back-stay cables. The results exhibit again the efficiency of the deck.

Another important point of the bridge is the base-section of the tower columns, which is first considered for the retrofit investigation for long-span bridges. The maximum value of the sectional forces including the axial force, the shear force and the bending moment is given in **Figure 13**. Due to noticeably high increase in the tensile axial force of the main and the

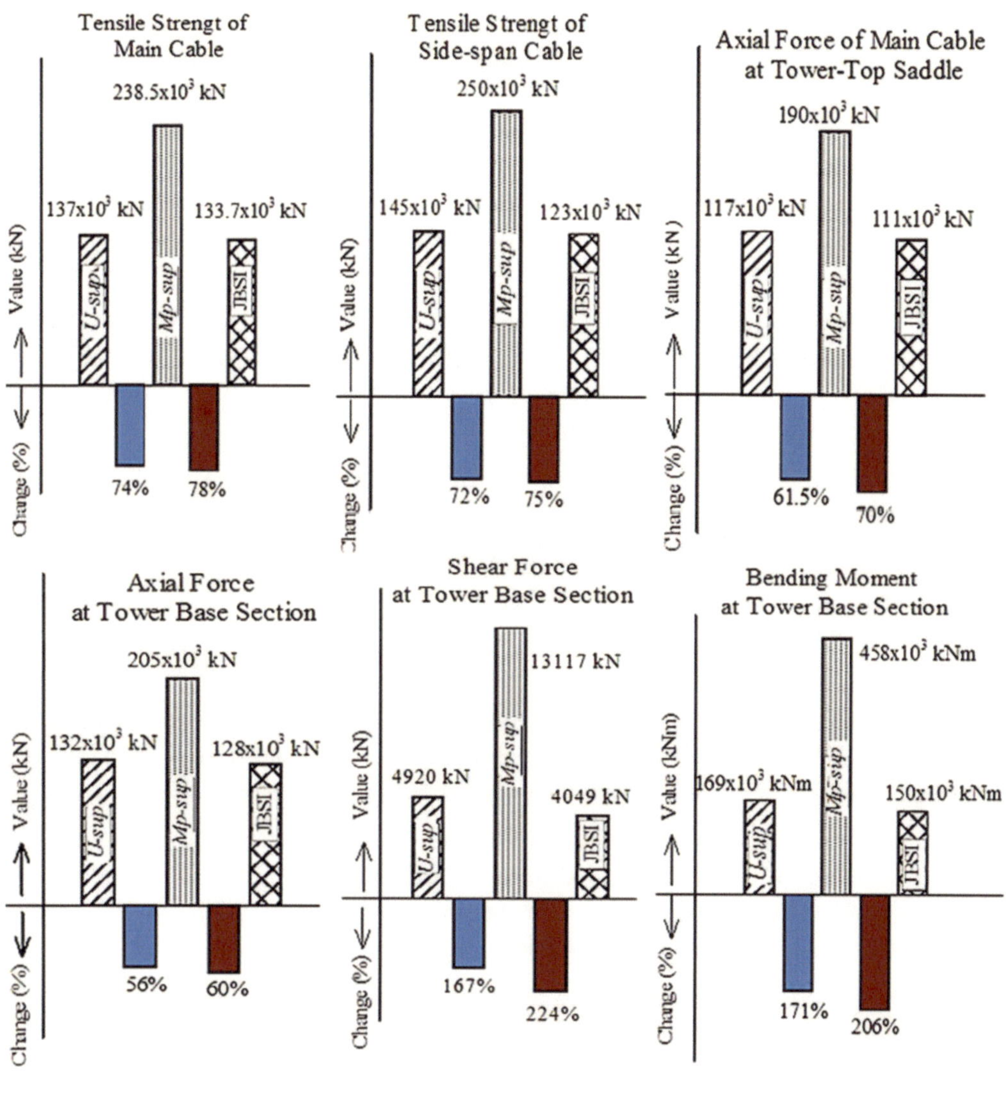

Figure 13. Comparison sectional force of the critical elements/points of the bridge.

back-stay cables, the axial force of the tower directly increased as 56% and 60% according to the U-sup and JBSI retrofit project, respectively. Although the shear force of the main cable at the tower top-saddle decreases, the shear force of the tower at the base considerably increases since the deck forces the tower at the level of the expansion joints (tower-deck connections) leading to high additional shear force. Therefore, the bending moment of the tower at the base increases highly as similar percentage increase to that of the shear force as shown in **Figure 13**.

6. Conclusion

In this study, St-Id concept is identified with implementation of the real suspension bridge, the Bosphorus Bridge. Each step of the concept from visualization to utilization is aimed to carry out for the bridge considering the strong wind event SHM data. The FEM calibration of the bridge is also made according to the outcomes from the experimental data analysis. Reliably established and calibrated FEM of the bridge is then adopted for the last step of St-Id approach, utilization/interpretation/decision-making. For this purpose, the multi-point earthquake analysis (Mp-sup) of the bridge is conducted. Based on the consequences of the study, the following important points are concluded;

- Under the extreme wind loading, the first modal frequency (lateral symmetric) of the bridge decreased in the range of 12%. The change was only obtained in the deck modes of the bridge. The tower modes are not affected from the strong wind. This is pertinent to high rigidity of tower box-section in transverse direction.
- After the extreme wind excitation, the natural frequencies of the bridge are also obtained, meaning that no damage is estimated under the strong wind. High modal periods during the wind indicated that damping of the bridge increased.
- The comparison between FEM and experimental results proved that the developed model could be reliably adopted for advanced structural analysis and performance prediction of the bridge.
- Under multi-point seismic excitation, the most critical components are determined as the main deck and tower base-section. The high increase in sectional force of the main and side-span cables is identified to be pertinent to the response of the main deck.
- In spite of no movement in the lateral direction at the expansion joints according to project specifications of the bridge, lateral movement is obtained from the Mp-sup analysis at these points. Hence, shear key elements are proposed to be considered to provide the restricted lateral movement at the expansion joints.

Based on the consequences obtained from this study, St-Id approach is shown to be a robust technique for overall assessment of long-span bridges. Apart from the implementation of the all steps of St-Id, each step also gives important results, which means that the concept can be divided into certain groups according to the aim of the study. Therefore, St-Id approach is recommended to be considered for performance prediction and structural condition evaluation of long-span cable-supported bridges.

Author details

Selcuk Bas

Address all correspondence to: selcukbas@itu.edu.tr

Department of Civil Engineering, Faculty of Engineering, Bartin University, Central Bartin, Turkey

References

[1] Necati Catbas F, Kijewski-Correa T, Emin Aktan A, editors. Structural Identification of Constructed Systems: Approaches, Methods, and Technologies for Effective Practice of St-Id. USA: American Society of Civil Engineers (ASCE); 2013

[2] Hart GC, Yao JTP. System identification in structural dynamics. Journal of the Engineering Mechanics Division. 1977;**103**(6):1089-1104

[3] Liu S-C, Yao JTP. Structural identification concept. Journal of the Structural Division. 1978;**104**(12):1845-1858

[4] Necati Catbas F, Ciloglu SK, Hasancebi O, Grimmelsman K, Aktan AE. Limitations in structural identification of large constructed structures. Journal of Structural Engineering. 2008;**133**(8):1051-1066

[5] Emin Aktan A, Yao JTP. On structural identification of constructed facilities. In: Building an International Community of Structural Engineers. USA: ASCE; 1996. p. 651-658

[6] Zhou Y, Zhang J, Yi W, Jiang Y, Pan Q. Structural identification of a concrete-filled steel tubular arch bridge via ambient vibration test data. Journal of Bridge Engineering. 2017;**22**(6):1-15. DOI: 10.1061/(ASCE)BE.1943-5592.0001086

[7] Sipple JD, Sanayei M. Full-scale bridge finite-element model calibration using measured frequency-response functions. Journal of Bridge Engineering. 2015;**20**(9):1-11. DOI: 10.1061/(ASCE)BE.1943-5592.0000705

[8] Kijewski T, Kareem A. Wavelet transforms for system identification in civil engineering. Computer-Aided Civil and Infrastructure Engineering. 2003;**18**(5):339-355. DOI: 10.1111/1467-8667.t01-1-00312

[9] Gul M, Necati Catbas F. Damage assessment with ambient vibration data using a novel time series analysis methodology. Journal of Structural Engineering. 2011;**137**(12):1518-1526

[10] Gokce HB, Necati Catbas F, Gul M, Frangopol DM. Structural identification for performance prediction considering uncertainties: Case study of a movable bridge. Journal of Structural Engineering. 2013;**139**(10):1703-1715

[11] Gul M, NecatiCatbas F. Structural health monitoring and damage assessment using a novel time series analysis methodology with sensor clustering. Journal of Sound and Vibration. 2011;**330**(6):1196-1210. DOI: 10.1016/j.jsv.2010.09.024

[12] Bas S, Apaydin NM, Ilki A, Necati Catbas F. Structural health monitoring system of the long-span bridges in Turkey. Structure and Infrastructure Engineering. 2017;**0**(0):1-20. Forthcoming. DOI: 10.1080/15732479.2017.1360365

[13] Chopra AK. Dynamics of Structures: Theory and Applications to Earthquake Engineering. 3rd ed. Pearson/Prentice Hall. NJ, US: Upper Saddle River; 2007. 876 p. DOI: 013156174X, 9780131561748

[14] Apaydin NM, Bas S, Harmandar E. Response of the Fatih Sultan Mehmet Suspension Bridge under spatially varying multi-point earthquake excitations. Soil Dynamics and Earthquake Engineering. 2016;**84**(5):44-54. DOI: 10.1016/j.soildyn.2016.01.018

[15] Bas S, Apaydin NM, Celep Z. Free vibration analysis and seismic performance assessment of two approach viaducts of Bosphorus Suspension Bridge. In: Caner A, Gülkan P, Mahmoud K, editors. Developments in International Bridge Engineering, Springer Tracts on Transportation and Traffic. Cham: Springer; 2016. p. 65-75. DOI: 10.1007/978-3-319-19785-2_6

[16] Bas S, Apaydin NM, Celep Z. Earthquake performance of the two approach viaducts of the Bosphorus Suspension Bridge. Earthquakes and Structures. 2016;**11**(3):387-406. DOI: 10.12989/eas.2016.11.3.387

[17] Freeman, Fox and Partners. Bosporus Bridge towers, suspended structures, cables, anchorage drawings. Consulting Engineers; Freeman, Fox and Partners, Westminster, London, SW.1; 1968

[18] Xu YL, Xia Y. Structural Health Monitoring of Long-Span Suspension Bridges. NY, US: CRC Press; March 31, 2017. 392 p. DOI: 9781138075634

[19] Emin Aktan A, Necati Catbas F, Grimmelsman KA, Pervizpour M. Development of a Model Health Monitoring Guide for Major Bridges. Drexel Intelligent Infrastructure and Transportation Safety Institute; Philadelphia, PA, US. 2003

[20] Apaydin NM, Dogan B, Zulfikar C. Chapter 92. Wind vibration characteristic of Bogazici suspension bridge using structural health monitoring data. In: Chen A, Frangopol DM, Ruan X, editors. Bridge Maintenance, Safety, Management and Life Extension. Abingdon, Oxon, UK: CRC Press; 2014. p. 703-707. DOI: 10.1201/b17063-102

[21] Apaydin NM, Zulfikar AC, Alcik H. Introduction of Bogazici suspension bridge structural health monitoring system. In: Proceedings of the 15th Word Conference on Earthquake Engineering, Lisbon; 2012

[22] The MathWorks Inc. MATLABR2015a: Commercial Integrating Technical Computing Program. Natick, MA, USA; 2014

[23] JBSI-Japanese Bridge and Structure Institute. Project Reports and Basic Design Documents for the project entitled "Seismic Reinforcement of Large Scale Bridges in Istanbul", prepared for the General Directorate of State Highways, Turkey; 2004

[24] Boore DM. Stochastic simulation of high-frequency ground motions based on seismological models of the radiated spectra. Bulletin of Seismological Society of America. 1983;**73**(6A):1865-1894

Chapter 5

Recent Advances in the Serviceability Assessment of Footbridges Under Pedestrian-Induced Vibrations

Javier Fernando Jiménez-Alonso and Andrés Sáez

Additional information is available at the end of the chapter

http://dx.doi.org/10.5772/intechopen.71888

Abstract

Current international guidelines determine the effect of pedestrians on footbridges via an equivalent harmonic load. However, the dynamic response of footbridges obtained according to these standards differs from the values recorded experimentally. In order to overcome this issue, a new modelling framework has been recently proposed by several researchers. This novel approach allows considering more accurately three key aspects: (i) the inter- and intra-subject variability, (ii) the pedestrian-structure interaction and (iii) the crowd dynamics. For this purpose, different crowd-structure interaction models have been developed. Despite the large number of proposals, all of them share the same scheme: the crowd-structure interaction is simulated by linking two sub-models, namely (i) a pedestrian-structure interaction sub-model and (ii) a crowd sub-model. Furthermore, the variability of the pedestrian's behaviour may be taken into account via the assumption that the model parameters are random variables. In this chapter, a summary of the state-of-art of this new modelling framework is presented, with special emphasis in a case study where the crowd-structure interaction model developed by the authors is used to simulate the lateral lock-in phenomenon on a real footbridge.

Keywords: footbridges, pedestrian-structure interaction, crowd dynamics, dynamic stability, vibration serviceability assessment

1. Introduction

The increase in the strength of the new structural materials together with the higher aesthetics requirements imposed by current modern society has led to the design of footbridges with greater slenderness, which may be prone to vibrate under pedestrian-induced excitations. There are three factors that characterize this engineering problem: (i) the vibration source, i.e. the pedestrian; (ii) the path, i.e. the structure; and (iii) the receiver, i.e. the pedestrian [1, 2]. In the last 20 years, after some vibratory events happened in several large-span footbridges [3–5], an intensive research activity has been conducted by the scientific community in order to better characterize the pedestrian-induced vibrations on footbridges. Concretely, these research efforts were mainly focused on two objectives: (i) the accurate definition of the vibration source [6] and (ii) the analysis of a remarkable event, the lateral lock-in phenomenon [7]. On the one hand, the determination of the load induced by pedestrians on footbridges was tackled progressively. Initially, the estimation of the force originated by a single walking or running pedestrian was studied [8, 9]. Subsequently, these results were further extrapolated to the case of a crowd moving on a footbridge [10]. On the other hand, the lateral lock-in instability phenomenon originated by the synchronization of a pedestrian flow walking on a footbridge has been widely studied as well. Based on the outcomes of these researches, different proposals to estimate the number of pedestrians that originates the lateral lock-in phenomenon, as well as limiting values of the modal properties of the structure to avoid the problem, have been provided [7].

As result of all these studies, several standards [11] and design guidelines [12] were published to facilitate designers the assessment of the vibration serviceability limit state of footbridges under pedestrian action. Although these design codes shed light on this issue, they still present some shortages, so that, the dynamic response of the structure obtained numerically based on these recommendations still differs from the values recorded experimentally [13].

In order to overcome these limitations, a new generation of models have been developed and proposed during the last 5 years, giving rise to a new modelling framework. Three key aspects have been additionally taken into account in order to improve the modelling of pedestrian flows and their effect on footbridges [14]: (i) the inter- and intra-subject variability of the pedestrian action, (ii) the pedestrian-structure interaction and (iii) the crowd behaviour. Furthermore, the variability of the pedestrian action is normally simulated via a probabilistic approach, considering that the parameters that characterize the crowd-structure interaction model may be defined as random variables [15]. All these proposed models share a common scheme, and the crowd-structure interaction is simulated via the linking of two sub-models [16, 17]: (i) a pedestrian-structure interaction sub-model and (ii) a crowd sub-model. For the pedestrian-structure sub-model, although different models have been proposed [18], the use of a single-degree-of-freedom (SDOF) system has gained wider popularity in the scientific community. For the crowd sub-model, two approaches have been proposed: either macroscopic or microscopic models [15]. In the first approach, the crowd behaviour is modelled based on fluid mechanics [10], whilse in the second, the position and velocity of each pedestrian follows a multi-agent law [19]. The second approach, which can account explicitly for the inter-subject variability of each pedestrian [20], has been internationally accepted as the best method to simulate numerically the behaviour of pedestrian flows [15]. The linking between the two sub-models is achieved by the implementation of several behavioural conditions [20]. In this way, if certain comfort limits are exceed by the pedestrian-structure interaction sub-model, the velocity and step frequency of each pedestrian in the crowd are modified [20, 21]. The new modelling framework, based on these crowd-structure interaction models, has been applied successfully to determine numerically the response of a footbridge under pedestrian action

[15–17], to study the change of the modal parameters of real footbridges under the effect of a group of pedestrians [22] and even to analyse the lateral lock-in phenomenon on real footbridges [23].

Nevertheless, despite all these advances, there is not currently any international design guideline which covers comprehensively all aspects of the problem, so it is a challenge for the next years to include all these research results in the design standards of such structures.

The chapter is organized as follows. First, some general recommendations on how to assess the vibration serviceability limit state of footbridges under pedestrian action, according to the more recent design guidelines, are presented in Section 2. Second, the main aspects of the new modelling framework to simulate the crowd-structure interaction are presented in Section 3. As this modelling framework divides the issue in two sub-models, in Section 4, the first sub-model, the pedestrian-structure sub-model is presented and in Section 5, the second sub-model, the crowd sub-model, is described. Later, the interaction between the two sub-models is explained and implemented in Section 6. Subsequently, a case study, the comparison of the analysis of the lateral lock-in phenomenon on the Pedro e Inês footbridge using three different approaches (the experimental values recorded during the field test, the numerical estimation according to the Synpex guidelines [12] and the new modelling framework) is presented in Section 7. The study shows the potential of this new modelling framework to assess more accurately the vibration serviceability limit state of footbridges under pedestrian action. Finally, Sections 8 and 9 present the main conclusions obtained from the chapter and future research lines to be explored, respectively.

2. Brief review of design standards

The international standards for the assessment of the vibration serviceability limit state of footbridges under pedestrian action share two general rules to tackle the pedestrian-induced vibration problem [6]: (i) the establishment of the range of frequencies that characterizes the pedestrian-structure interaction (**Table 1**) and (ii) the treatment of the problem separately in terms of the direction in which the pedestrian action (longitudinal, lateral or vertical) is applied. However, most of these standards only establish the need to assess the dynamic behaviour of the structure, if some of its natural frequencies are within the interaction range (**Table 1**), but do not define a methodology to check the required comfort level.

According to the authors' opinion, the Synpex guidelines [12] are currently the most comprehensive standard to assess the vibration serviceability limit state of footbridges under pedestrian action. These guidelines [12] divide the checking of the vibration serviceability limit state in seven steps:

i. Evaluate, numerically, the natural frequencies of the footbridge based on a finite element model of the structure.

ii. If some of the natural frequencies of the structure lie inside the interaction ranges (**Table 1**), the comfort class of the footbridge must be further checked.

Standards	Vertical [Hz]	Lateral [Hz]
LRFD American Guide (2009)	<3.00	
Eurocode 1 (2002)	1.60-2.40	0.80–1.20
Eurocode 5 (2003)	<2.50	0.80–1.20
DIN-Fachbericht 102 (2003)	1.60–2.40/3.50–4.50	
SIA 260 (2003)	1.60–4.50	<1.30
BS 5400 (2006)	<5.00	<1.50
Austroads (2012)	1.50–3.00	
Hong Kong Guide (2009)	1.50–2.30	
Ontario Guide (1995)	<3.00	
Setra (2006)	1.00–2.60/2.60–5.00	0.30–1.30/1.30–2.50
Synpex (2007)	1.25–2.30/2.50–4.60	0.50–1.20
EHE-08 (2008)	<5.00	
EAE (2011)	1.60–2.40/3.50–4.50	0.60–1.20
IAP-11 (2011)	1.25–4.60	0.50–1.20

Table 1. Ranges of frequencies of pedestrian-structure interaction according to different international standards [14].

iii. Different design scenarios must be assessed: for each design scenario, the expected traffic class in terms of the pedestrian density, d [P=Person/m^2], (**Table 2**) and its corresponding comfort class in terms of limit acceleration (**Table 3**) must be determined according to the owner's requirements.

Class	Density d [P/m^2]	Characteristics
TC1	<15 P (P = Person)	Very weak traffic
TC2	<0.20 P/m^2	Comfortable and free walking
TC3	<0.50 P/m^2	Unrestricted walking and significantly dense traffic
TC4	<1.00 P/m^2	Uncomfortable situation and obstructed walking
TC5	<1.50 P/m^2	Unpleasant walking and very dense traffic

Table 2. Traffic classes [12].

Class	Degree	Vertical acceleration	Horizontal acceleration
CL1	Maximum	<0.50 m/s^2	<0.10 m/s^2
CL2	Medium	0.50–1.00 m/s^2	0.10–0.30 m/s^2
CL3	Minimum	1.00–2.50 m/s^2	0.30–0.80 m/s^2
CL4	Discomfort	>2.50 m/s^2	>0.80 m/s^2

Table 3. Defined comfort classes with limit acceleration ranges [12].

iv. The damping ratio of the affected vibration mode, ζ_f, is estimated in function of the construction type and the amplitude of the vibrations [12].

v. The maximum acceleration has to be evaluated for each design scenario. For this purpose, it is necessary to define a load model which may be characterized by the following equivalent harmonic loads [12]:

- A pedestrian stream walking is simulated by an equivalent load:

$$p_{wal}(t) = G \cdot \cos\left(2 \cdot \pi \cdot f_s \cdot t\right) \cdot n' \cdot \psi / L_f \ [\mathrm{N/m}] \quad (1)$$

- A pedestrian jogging is simulated by a single vertical moving load:

$$P_{jog}(t, v_p) = 1250 \cdot \cos\left(2 \cdot \pi \cdot f_s \cdot t\right) \cdot \psi \ [\mathrm{N}] \quad (2)$$

where $G \cdot \cos\left(2 \cdot \pi \cdot f_s \cdot t\right)$ is the harmonic load due to a single pedestrian, with G being the dynamic load factor (DLF) of the pedestrian step load (280 N for vertical direction, 140 N for longitudinal direction and 35 N for lateral direction); f_s is the step frequency [Hz], which is assumed equal to the considered natural frequency, f_f; ψ is the reduction coefficient that takes into account the probability that the footfall frequency approaches the considered natural frequency and it may be estimated from **Figure 1**, according to the considered natural frequency; v_p is the pedestrian velocity [m/s] which may be assumed around 3 m/s [12] and L_f is the length of the footbridge [m]. In Eq. (1), n' is the equivalent number of pedestrians on the footbridge, which may be determined from:

$$n' = \begin{matrix} 10.8 \cdot \sqrt{\zeta_f \cdot n} \\ 1.85 \cdot \sqrt{n} \end{matrix} \ \text{if} \ \begin{matrix} d < 1.00 \ P/m^2 \\ d \geq 1.00 \ P/m^2 \end{matrix} \quad (3)$$

in terms of the number of pedestrians on the deck, n, and the damping ratio of the considered vibration mode, ζ_f.

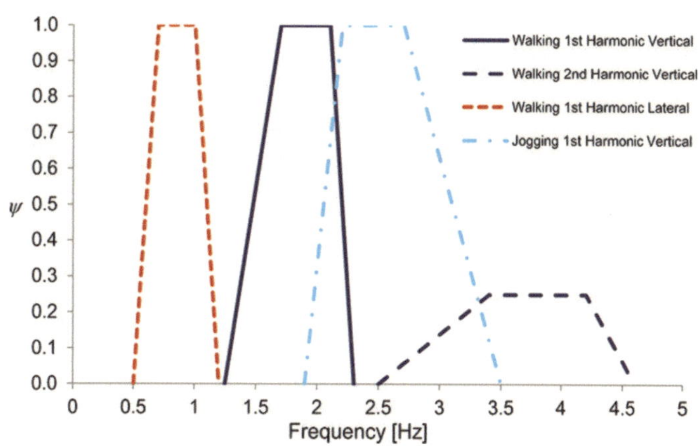

Figure 1. Pedestrian reduction coefficient, ψ, for the equivalent pedestrian load [12].

To estimate the considered natural frequency for each design scenario, the mass of pedestrians has to be taken into account (with a medium pedestrian weight about 70 kg) when its value is greater than 5% of the modal deck mass.

vi. The dynamic response obtained for each considered design scenario must be compared with the trigger acceleration amplitude, 0.10–0.15 m/s^2, which avoids the occurrence of the lateral lock-in phenomenon.

vii. The estimated dynamic acceleration is then compared with the specified comfort class. In case of non-compliance, the designer must adopt measures to improve the dynamic behaviour of the structure, such as for instance: (i) the modification of the mass of the deck, (ii) the modification of the natural frequencies of the structure and/or (iii) the increase of the damping [12].

In spite of the fact that the Synpex design guidelines [12] were an important breakthrough, they still present several limitations, which originate that the numerical prediction of the dynamic response of footbridges, obtained using them, under- or over-estimates the values recorded experimentally. As main limitations, the following ones may be enumerated: (i) the change of the dynamic properties of the structure, due to the presence of pedestrians, is estimated in a simplified form, adding directly the pedestrian mass to the structural mass without considering any additional effect on the remaining modal parameters of the structure, (ii) the proposed methods do not fit well to the case where several vibration modes of the footbridge are affected by the pedestrian-induced excitations, (iii) the effect of the non-synchronized pedestrians are not taken into account by these recommendations and (iv) the definition of the pedestrian load is performed under a deterministic approach which does not allow considering the inter- and intra-subject variability of the pedestrian action. In order to overcome these limitations, a new generation of models that configure a new modelling framework has been proposed. A brief description of this new modelling framework is included in the next section.

3. New modelling framework of crowd-structure interaction

The most recent research on this topic proposes and further implements several crowd-structure interaction models to better characterize the dynamic response of footbridges under pedestrian action [14–17]. All these models, which share a common scheme, constitute a new modelling framework for this engineering problem. According to this new approach, the crowd-structure interaction is simulated by linking two individual sub-models (**Figure 2**): (i) a pedestrian-structure interaction sub-model and (ii) a crowd sub-model.

In the first sub-model, although there are several proposals [18, 22, 24, 25] to simulate the pedestrian action (single-degree-of-freedom (SDOF) system, multiple degrees of freedom (MDOF) system and inverted pendulum (IP) system), the most widely adopted alternative is to model the pedestrian either as a SDOF system in vertical direction [18] or as a IP system in lateral direction [20, 24], while the structure is characterized via its modal parameters [22, 26].

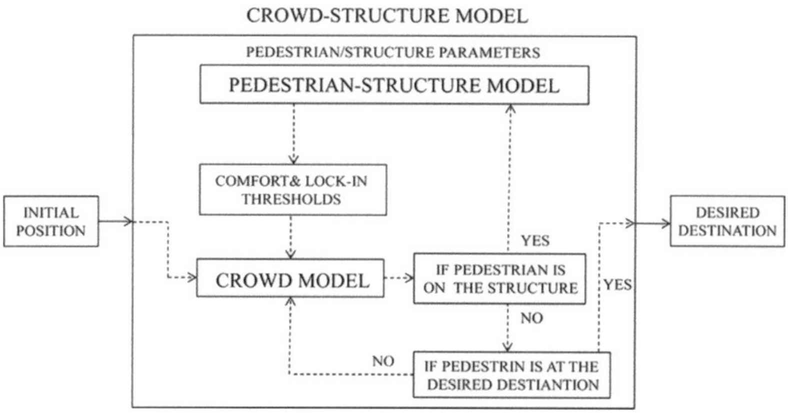

Figure 2. Layout of the new modelling framework.

All the pedestrian-structure interaction models based on the use of a SDOF system share a common formulation to solve the pedestrian-structure interaction [22, 26] but, however, they differ in the values adopted to characterize the modal parameters of the SDOF systems. A wide summary of the pedestrian-structure interaction models proposed by different authors can be found in Ref. [18]. The main output obtained from this sub-model is usually the acceleration experienced by each pedestrian.

In the second sub-model, the crowd is usually simulated via a behavioural model [19] that provides a description of the individual pedestrian position, $\mathbf{x}_p$, pedestrian velocity, $\mathbf{v}_p$, and step pedestrian frequency, f_s. Additionally, in order to take into account the synchronization among pedestrians, an additional parameter must be included. A common manner to simulate this phenomenon is to add a different phase shift, ϕ_p, in the definition of the ground reaction load generated by each pedestrian [14].

The linking between the two sub-models is usually achieved in the different proposals by taking into account the modification of the pedestrian behaviour in terms of the vibration level that he/she experiences [15, 17, 20–24]. Two additional conditions are commonly included for this purpose: (i) a retardation factor, which reduces the pedestrian velocity in terms of the accelerations experienced by each pedestrian; and (ii) a lateral lock-in threshold, which allows simulating the synchronization among the pedestrians and the structure by the modification of both their step frequencies and the phases [20–23]. This new approach has only been implemented, to the best of the authors' knowledge, in vertical and lateral direction, since there are few reported cases of pedestrian-induced vibration problems in longitudinal direction. In order to illustrate briefly this new modelling framework, one of the most recent crowd-structure interaction models, which has been proposed by the authors, is described in the next sections [23]. Subsequently, the potential of the approach to accurately assess the vibration serviceability limit state of footbridges under pedestrian action is illustrated via its implementation for the analysis of a case study. For clarity, the model is described here only for the lateral direction, although it may be easily generalized to the vertical direction [14].

4. Modelling pedestrian-structure interaction

The pedestrian-structure interaction model may follow from the application of dynamic equilibrium equations between a SDOF-system (**Figure 3a**) and the footbridge (**Figure 3b**). The pedestrian mass is divided into sprung, m_a, and unsprung, m_s, components [kg].

As result of this dynamic equilibrium, the following coupled equation system may be obtained [22]:

$$M_i\ddot{y}_i + C_i\dot{y}_i + K_i y_i = \phi_{num\_i}(x_p) \cdot F_{\text{int}} \tag{4}$$

$$m_a\ddot{y}_a + c_p(\dot{y}_a - \dot{y}_s) + k_p(y_a - y_s) = 0 \tag{5}$$

$$m_s\ddot{y}_s + c_p(\dot{y}_s - \dot{y}_a) + k_p(y_s - y_a) = F_p - F_{\text{int}} \tag{6}$$

where y_i is the amplitude of the vibration mode *ith* of the footbridge [m]; y_a is the displacement of the pedestrian sprung mass [m]; y_s is the displacement of the pedestrian unsprung mass [m]; k_p is the pedestrian stiffness [N/m]; c_p is the pedestrian damping [sN/m]; F_p is the ground reaction force [N]; F_{int} is the pedestrian-structure interaction force [N]; M_i is the mass associated with the *ith* vibration mode [kg]; C_i is the damping associated with the *ith* vibration mode [sN/m]; K_i is the stiffness associated with the *ith* vibration mode [N/m]; ϕ_{num_i} is the modal coordinates of the *ith* vibration mode; $x_p = v_{px} \cdot t$ is the pedestrian's longitudinal position on the footbridge [m], being t the time [sec.] and v_{px} the longitudinal component of the pedestrian velocity vector [m/s]; d_p is the distance among pedestrians [m] and $w(x,t)$ is the lateral displacement of the footbridge at position x [m].

The numerical vibration modes, $\phi_{num\_i}(x)$, may be obtained by a numerical modal analysis of the structure based on the finite element method:

$$\phi_{num\_i}(x) = \sum_j \phi_i^j \cdot N_j(x) \tag{7}$$

where $N_j(x)$ is the beam shape functions and ϕ_i^j is the nodal values of the vibration modes.

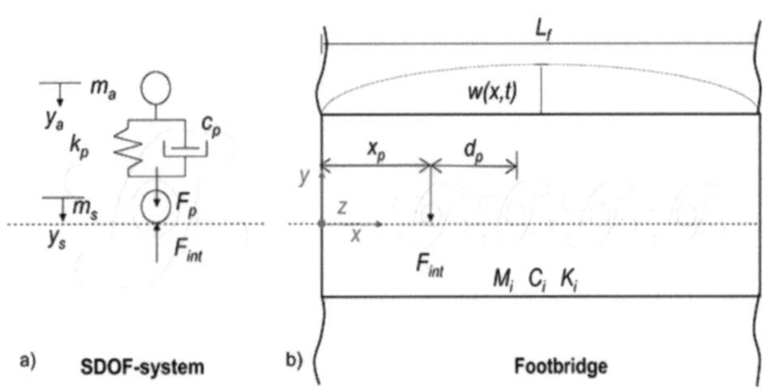

Figure 3. Biomechanical pedestrian-structure interaction model in lateral direction [14]. (a) SDOF-system and (b) Footbridge.

Thus, F_{int} follows from the above Eq. (6) to yield:

$$F_{\text{int}} = F_p - m_s\ddot{y}_s - c_p(\dot{y}_s - \dot{y}_a) - k_p(y_s - y_a) \tag{8}$$

and substituting this equation into the equilibrium equation of the structure:

$$M_i\ddot{y}_i + C_i\dot{y}_i + K_iy_i = \phi_{num\_i}(x_p)\cdot(F_p - m_s\ddot{y}_s - c_p(\dot{y}_s - \dot{y}_a) - k_p(y_s - y_a)) \tag{9}$$

Applying, at the contact point, the equations of compatibility of displacements, $y_s = w(x_p, t)$, velocities, $\dot{y}_s = \dot{w}(x_p, t)$ and accelerations, $\ddot{y}_s = \ddot{w}(x_p, t)$, between the SDOF system and the structure, the following expressions may be obtained:

$$w(x_p, t) = \sum_{i=1}^{n_m} y_i(t)\cdot\phi_{num\_i}(x_p) \tag{10}$$

$$\dot{w}(x_p, t) = \sum_{i=1}^{n_m}\dot{y}_i(t)\cdot\phi_{num\_i}(x_p) + \sum_{i=1}^{n_m} y_i(t)\cdot v_{px}\cdot\phi'_{num\_i}(x_p) \tag{11}$$

$$\ddot{w}(x_p, t) = \sum_{i=1}^{n_m}\ddot{y}_i(t)\cdot\phi_{num\_i}(x_p) + \sum_{i=1}^{n_m} 2\cdot\dot{y}_i(t)\cdot v_{p,x}\cdot\phi'_{num\_i}(x_p) + \sum_{i=1}^{n_m} y_i(t)\cdot {v_{p,x}}^2\cdot\phi''_{num\_i}(x_p)) \tag{12}$$

where $\phi'_{num\_i}(x)$ and $\phi''_{num\_i}(x)$ the first and second spatial derivatives of the *ith* numerical vibration mode and n_m, is the number of considered vibration modes.

It is assumed that the lateral displacement of the footbridge may be decomposed in terms of the amplitude $y_i(t)$ and the modal coordinates of the n_m vibration modes, $\phi_{num\_i}(x)$, and the time variation of the pedestrian velocity is neglected due to its low contribution: Subsequently, the above relations Eqs. (10)–(12) may be substituted in the overall dynamic equilibrium equation of the footbridge, obtaining the following pedestrian-structure interaction model equations to yield in matrix form:

$$\mathbf{M}(t)\cdot\ddot{\mathbf{y}}(t) + \mathbf{C}(t)\cdot\dot{\mathbf{y}}(t) + \mathbf{K}(t)\cdot\mathbf{y}(t) = \mathbf{F}(t) \tag{13}$$

In this manner, the pedestrian-structure interaction model may be represented by a system with $(n_m + 1)$ equations (being n_m the number of the considered vibration modes and *1* the SDOF system that simulates the pedestrian behaviour). In case of a group of *k* pedestrians (**Figure 3b**), the number of equations of system increases to $(n_m + k)$, maintaining the same scheme. A more detailed description of this pedestrian-structure interaction model may be found in Ref. [22].

The lateral ground reaction force, F_p, generated by each pedestrian, may be defined under either a deterministic [8] or a stochastic approach [15]. The second approach allows taking into account the inter- and intra-subject variability of the pedestrian action [15]. Although there are more complex ways [15] to define the lateral ground reaction force, however, it is usually expressed in terms of a Fourier series decomposition [8, 9, 12] as:

$$F_p = m \cdot g \sum_{i=1}^{n_f} \alpha_i \cdot \sin\left(\pi \cdot i \cdot f_s \cdot t - \varphi_i - \phi_p\right) \tag{14}$$

where $m = m_a + m_s$ is the total pedestrian mass, g is the acceleration of gravity, α_i is the Fourier coefficients of the *ith* harmonic of the lateral force, f_s [Hz] is the pedestrian step frequency, φ_i is the phase shifts of the *ith* harmonic of the lateral pedestrian force, ϕ_p is the phase shift among pedestrians and n_f is the total number of contributing harmonics.

According to this formulation, the deterministic or stochastic character of the pedestrian-structure interaction sub-model can be considered depending on the way in which the parameters of the model are defined. If a fixed value is assigned to the parameters, the sub-model will be deterministic; however, if the parameters are defined as random variables, the sub-model will be stochastic.

Finally, **Table 4** shows the values reported in Ref. [14] for the characterization of the pedestrian-structure interaction model. Additionally, a wide summary of the parameters proposed by other researchers can be found in Ref. [18]. These values (**Table 4**) allow defining the pedestrian-structure interaction model in either a deterministic form (considering the average values) or a stochastic form (considering the probabilistic distribution), depending on the purpose of the case under study.

	Pedestrian modal parameters	
	Lateral	
Definition	**Parameter**	**Value**
Pedestrian total mass	m	$N(75, 15)$kg
Pedestrian sprung mass	m_a	$N(73.216, 2.736)$%
Pedestrian damping ratio	ζ_p	$N(49.116, 5.405)$%
Pedestrian natural frequency	f_p	$N(1.201, 0.178)$Hz
	Walking pedestrian force	
	Lateral	
Definition	**Parameter**	**Value**
First harmonic	α_1	$N(0.086, 0.017)$
Second harmonic	α_2	$N(0.094, 0.009)$
Third harmonic	α_3	$N(0.040, 0.019)$
First phase shift	φ_1	0°
Second phase shift	φ_2	0°
Third phase shift	φ_3	0°

Table 4. Parameters of the pedestrian-structure interaction sub-model reported in Refs. [14, 15], where $N(\mu, \sigma)$ is a Gaussian distribution with mean value, μ, and the standard deviation, σ.

http://dx.doi.org/10.5772/intechopen.71888

5. Modelling crowd dynamics

The pedestrian moving inside a crowd may be modelled using either a macroscopic [10] or a microscopic model [15]. The second option is currently the most utilized and it has been successfully implemented by several authors [15, 19–22]. According to this approach, the movement of each pedestrian is governed by the dynamic balance among particles [14]. This model assumes that the different motivations and influences experimented by the pedestrians are described by different social forces [19]. The model is based on Newton dynamics and is able to represent the following rules in relation with the natural pedestrian movement (see Ref. [19] for a more detailed description): (i) the fastest route is usually chosen by pedestrians, (ii) the individual speed of each pedestrian follows a probabilistic distribution function and (iii) the distance among pedestrians in a crowd depends on the pedestrian density, the spatial configuration of the crowd and the pedestrian speed. As an example, the different social forces acting between two pedestrians in a crowd are illustrated in **Figure 4**.

In this manner, the multi-agent model that simulates the behaviour of the crowd consists of the sum of three partial forces: (i) the driving force, $\mathbf{F}_{dri}$, (ii) the repulsive force generated by the interaction among pedestrians, $\mathbf{F}_{ped}$, and (iii) the repulsive force generated by the interaction with the boundaries, $\mathbf{F}_{bou}$. A detailed description of these three forces is carried out in the next sub-sections. The sum of these three forces generates the overall pedestrian-crowd interaction

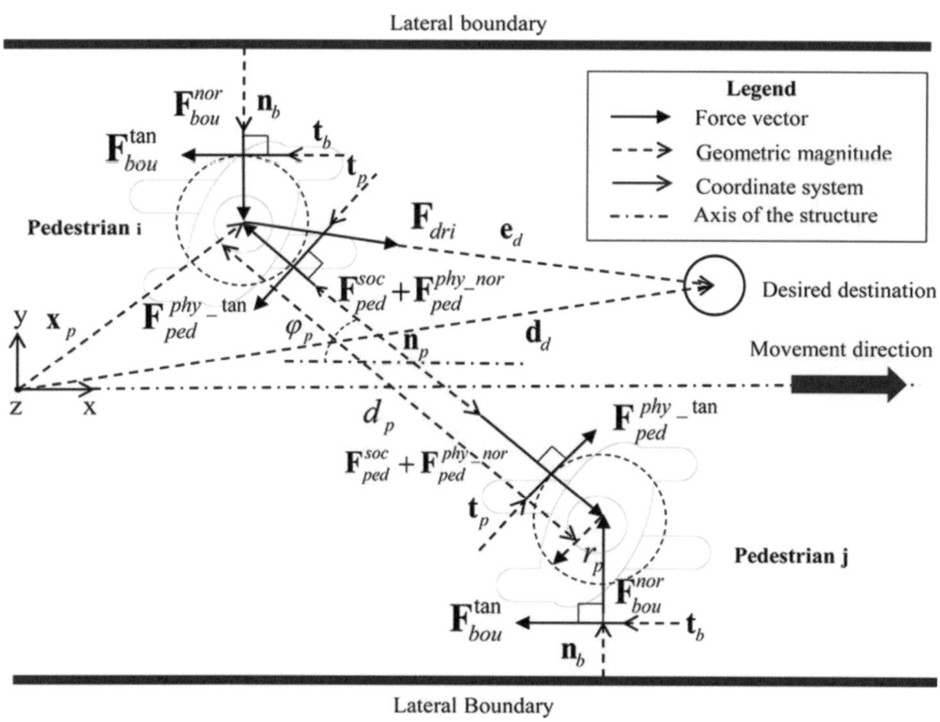

Figure 4. Biomechanical pedestrian-structure interaction model [14].

force, $\mathbf{F}_{pci}$, that describes the movement and direction of each pedestrian in the crowd. This resultant force is defined as follows:

$$\mathbf{F}_{pci} = \mathbf{F}_{dri} + \mathbf{F}_{ped} + \mathbf{F}_{bou} \tag{15}$$

5.1. Driving force

Each pedestrian has a certain motivation to reach his/her desired destination [19], $\mathbf{d}_d$, with his/her desired velocity, v_d, which is represented by the driving force, $\mathbf{F}_{dri}$, as:

$$\mathbf{F}_{dri} = m \cdot \left(\frac{v_d \cdot \mathbf{e}_d}{t_r} - \frac{\mathbf{v}_p}{t_r} \right) \tag{16}$$

where $\mathbf{e}_d$ is the desired direction vector, $\mathbf{v}_p$ is the pedestrian step velocity and t_r is the relaxation time (the time it takes a pedestrian to adapt its motion to its preferences).

5.2. Interactions among pedestrians

The interaction among pedestrians originates a repulsive force [19], $\mathbf{F}_{ped}$, with two components, a socio-psychological force, $\mathbf{F}_{ped}^{soc}$, and a physical interaction force, $\mathbf{F}_{ped}^{phy}$, as:

$$\mathbf{F}_{ped} = \mathbf{F}_{ped}^{soc} + \mathbf{F}_{ped}^{phy} \tag{17}$$

The socio-psychological force reflects the fact that the pedestrians try to maintain a certain distance to other pedestrians in the crowd. This socio-psychological force depends on the distance between pedestrians, reaching its maximum value at the initial distance between two pedestrians, d_p, and tending to zero as such distance increases. The socio-psychological force is defined as:

$$\mathbf{F}_{ped}^{soc} = A_p \cdot \exp\left(\frac{2 \cdot r_p - d_p}{B_p} \right) \cdot \mathbf{n}_p \cdot s_p \tag{18}$$

where A_p is the interaction strength between two pedestrians; B_p is the repulsive interaction range between pedestrians; r_p is the so-called pedestrian radius; $\mathbf{n}_p$ is the normalized vector pointing between pedestrians and s_p is a form factor to consider the anisotropic pedestrian behaviour [19], whose value may be obtained from:

$$s_p = \lambda_p + \left(1 - \lambda_p\right) \cdot \frac{1 + \cos\left(\varphi_p\right)}{2} \tag{19}$$

being, λ_p, a coefficient that takes into account the influence of the pedestrians placed in front of the subject on his/her movement, and, φ_p, the angle formed between two pedestrians.

In situations of physical contact among pedestrians ($d_p \leq 2 \cdot r_p$) and high pedestrian density ($\geq$0.80 P=Person/m^2), the physical interaction force, $\mathbf{F}_{ped}^{phy}$, must be considered. This force may be divided in other two components: (i) the body force, $\mathbf{F}_{ped}^{phy\_nor}$, and (ii) the sliding force, $\mathbf{F}_{ped}^{phy\_\tan}$. The first component simulates the counteracting body action that each pedestrian performs to avoid physical damage in case he/she gets in physical contact with other individuals. The second component considers the pedestrians' tendency to avoid overtaking other subjects quickly at small distances [19]. It is defined as [22]:

$$\mathbf{F}_{ped}^{phy} = \mathbf{F}_{ped}^{phy\_nor} + \mathbf{F}_{ped}^{phy\_\tan} \tag{20}$$

$$\mathbf{F}_{ped}^{phy\_nor} = C_p \cdot H(2 \cdot r_p - d_p) \cdot \mathbf{n}_p \tag{21}$$

$$\mathbf{F}_{ped}^{phy\_\tan} = D_p \cdot H(2 \cdot r_p - d_p) \cdot \Delta v_p^t \cdot \mathbf{t}_p \tag{22}$$

being $\mathbf{F}_{ped}^{phy\_nor}$ the normal component of the physical interaction force; $\mathbf{F}_{ped}^{phy\_\tan}$ the tangential component of the physical interaction force; C_p the body force strength due to the contact between pedestrians; D_p the sliding force strength due to the contact between pedestrians; $\mathbf{t}_p$ a normalized tangential vector (which is perpendicular to $\mathbf{n}_p$); $\Delta v_p^t = \langle \Delta \mathbf{v}_p \cdot \mathbf{t}_p \rangle$ the component of the relative pedestrian velocity in tangential direction; $\Delta \mathbf{v}_p$ the vector of differential velocities between two pedestrians; and $H(\bullet)$ a function which may be defined as [22]:

$$H(\bullet) \cdot = \begin{cases} \bullet & if \quad \bullet > 0 \\ 0 & if \quad \bullet \leq 0 \end{cases} \tag{23}$$

5.3. Interactions with boundaries

The interaction with the boundaries gives rise to forces, $\mathbf{F}_{bou}$. These forces are equivalent to the ones resulting from the interaction with other pedestrian, so they can be formulated in a similar fashion as [22]:

$$\mathbf{F}_{bou} = \mathbf{F}_{bou}^{nor} + \mathbf{F}_{bou}^{\tan} \tag{24}$$

$$\mathbf{F}_{bou}^{nor} = \left\{ A_b \cdot \exp\left(\frac{r_p - d_b}{B_b}\right) + C_b \cdot H(r_p - d_b) \right\} \cdot \mathbf{n}_b \tag{25}$$

$$\mathbf{F}_{bou}^{\tan} = D_b \cdot H(r_p - d_b) \cdot \langle \mathbf{v}_p, \mathbf{t}_b \rangle \cdot \mathbf{t}_b \tag{26}$$

being $\mathbf{F}_{bou}^{nor}$ the component of the boundary interaction force in normal direction; $\mathbf{F}_{bou}^{\tan}$ the component of the boundary interaction force in tangential direction; A_b the pedestrian-boundary interaction strength; B_b the pedestrian-boundary repulsive interaction range; d_b the pedestrian-boundary distance; C_b the body force strength due to the contact with the boundary; D_b the sliding force strength due to the contact with the boundary; $\mathbf{n}_b$ the normalized normal vector between the pedestrian and boundary; $\mathbf{t}_b$ the normalized tangential vector (which is perpendicular to $\mathbf{n}_b$) and $\langle\rangle$ denotes the scalar product [22].

All the parameters for the considered crowd sub-model, based on the social force model, may be obtained from the results provided by different authors [19, 20] as summarized in **Table 5**.

Parameter	Element	Value
Relaxation time	t_r	0.50 sec.
Interaction strength pedestrians	A_p	2000 N
Interaction range pedestrians	B_p	0.30 m
Potential factor	λ_p	0.20
Contact strength pedestrians	C_p	2000 N
Sliding strength pedestrians	D_p	4800 N
Interaction strength boundaries	A_b	5100 N
Interaction range boundaries	B_b	0.50 m
Contact strength boundaries	C_b	2000 N
Sliding strength boundaries	D_b	4800 N
Radius of pedestrian	r_p	0.20 m

Table 5. Parameters of the crowd sub-model reported in Refs. [19, 20].

5.4. Simulation procedure

The simulation of a pedestrian flow requires the determination of four parameters: (i) the pedestrian density, d, (ii) the desired velocity of each pedestrian, v_d, (iii) the phase shift among pedestrians, ϕ_p, and (iv) the distance among pedestrians, d_p.

First, the pedestrian density, d, is established according to the owner's requirements [12]. Second, the values of the desired velocity of each pedestrian can be obtained from the pedestrian step frequencies, f_s, assuming that initially the pedestrian velocity, v_p, is equal to the desired velocity, v_d. For this purpose, the Gaussian distribution of the pedestrian step frequency, $N(1.87, 0.186)$ Hz, reported in Ref. [2], can be adopted as reference. After assigning a step frequency to each pedestrian, its desired velocity is determined from the empirical relation given in Ref. [27]:

$$f_s = 0.35 \cdot |\mathbf{v}_p|^3 - 1.59 \cdot |\mathbf{v}_p|^2 + 2.93 \cdot |\mathbf{v}_p| \tag{27}$$

Subsequently, the initial phase shift among pedestrians, ϕ_p, which allows estimating the number of pedestrian that arrive at the footbridge in phase, is determined considering that it follows a Poisson distribution [14]. Finally, the original distance among pedestrians is calculated considering the width of the footbridge, a predefined geometrical-shaped mesh of pedestrians (triangular or rectangular) and the considered pedestrian density.

The acceleration vector, $\mathbf{a}_p$, that acts on each pedestrian may be determined as:

$$\mathbf{a}_p = \frac{\mathbf{F}_{pci}}{m} \tag{28}$$

Finally, the evaluation of the remaining variables that govern the crowd model, $\mathbf{v}_p$ and $\mathbf{x}_p$, are then performed using a multi-step method [14].

6. Modelling crowd-structure interaction

The crowd-structure interaction is usually modelled including additional behavioural conditions [20–22]. Concretely, two requirements have been included in this proposal: (i) a comfort and (ii) a lateral lock-in threshold [14, 20].

First, a comfort condition is usually included in the crowd-structure interaction model to take into account the modification of the behaviour of each pedestrian due to the change of his/her comfort level. For this purpose, a retardation factor has been applied to the pedestrian velocity. A minimum comfort threshold 0.20 m/s^2 is selected following the results provided by several researches [20, 28]. In this manner, if the lateral acceleration of each pedestrian, $\ddot{y}_a$, is above this value, the pedestrian velocity is reduced by a retardation factor, r_v, which is a function of the acceleration experienced by the pedestrian. Following the intuitive assumption, reported in Ref. [20], that the pedestrians are likely to react more firmly as the lateral acceleration they feel is higher, a tri-linear function is considered, Eq. (29).

$$r_v(\ddot{y}_a) = \begin{matrix} 1 - (0.1/1.05) \cdot \ddot{y}_a \\ 0.9 - (0.3/0.65) \cdot (\ddot{y}_a - 1.05) \\ 0.6 - (0.6/0.4) \cdot (\ddot{y}_a - 1.7) \\ 0 \end{matrix} \quad \text{if} \quad \begin{matrix} \ddot{y}_a \leq 1.05\ m/s^2 \\ \ddot{y}_a \leq 1.7\ m/s^2 \\ \ddot{y}_a \leq 2.1\ m/s^2 \\ \ddot{y}_a > 2.1\ m/s^2 \end{matrix} \tag{29}$$

On the other hand, a maximum lateral limit acceleration, $\ddot{y}_{\text{lim}} = 2.10$ m/s^2, have also been considered [29], so pedestrians stop walking, when the experienced acceleration becomes too high, to keep their balance, and they remain stopped until the footbridge reduces its accelerations. Both to stop walking and to remain stationary before starting to walk again, the same reaction time, $t_{rea} = 2.00$ s, has been adopted. A linear variation has been considered to simulate the variation of the pedestrian velocity during the reaction time. Additionally, a practical lower limit of the pedestrian velocity has been established in order to avoid meaningless small values of this magnitude [20].

$$|v_p| = \begin{cases} 0.1 \cdot |v_d| & \text{if} \quad \ddot{y}_a < \ddot{y}_{\text{lim}} \cap |v_p| \leq 0.1 \cdot |v_d| \\ 0 & \text{if} \quad \ddot{y}_a \geq \ddot{y}_{\text{lim}} \end{cases} \tag{30}$$

Finally, as lateral lock-in threshold, the criterion suggested by the French standard [11] is usually adopted to simulate the synchronization phenomenon between the movement of the crowd and the footbridge. For this purpose, both the step frequency, f_s, and phase shift, ϕ_p, of each pedestrian are modified to match the natural frequency of the structure, if the lateral acceleration experienced by each subject is above 0.15 m/s^2 and its step frequency is within $\pm$10% of the lateral natural frequency of the structure [30].

7. Application example: Lateral lock-in phenomenon on a real footbridge

In order to illustrate the potential of this new modelling framework, the analysis of the lateral lock-in phenomenon on a real footbridge, the Pedro e Inês footbridge (Coimbra Portugal) has been performed [23]. The maximum lateral accelerations at the mid-span of the footbridge obtained via three different methods during a lateral lock-in pedestrian test are correlated. The three method used are: (i) the experimental values recorded during a lateral lock-in pedestrian test reported in Ref. [31], (ii) the numerical estimation of the maximum lateral acceleration obtained according to the Synpex guidelines [12] and (iii) the numerical prediction obtained based on the application of the proposed approach [23]. On an updated finite element model of the structure [32].

The footbridge is situated over the Mondego River at Coimbra (Portugal). The structure is configured by five spans (total length of 274.5 m); a central arch of 110 m, two lateral semi-arches of 64 m and two transition spans of 30.5 and 6 m, respectively (**Figure 5**). The deck is configured by a concrete-steel composite box-girder with a variable width between 4 and 8 m. The footbridge presents an anti-symmetrical configuration with respect to the longitudinal axis of the structure. In this way, the intersection of the two parallel decks generates a panoramic square at mid-span of the footbridge (**Figure 5**). As result of the numerical studies performed during the design phase, it was checked that the structure was prone to pedestrian-induced vibrations in lateral direction. Experimental tests were conducted to assess the dynamic response of the footbridge under pedestrian action in lateral direction. The main outcomes of this experimental work were reported in Ref. [31]. These results have been employed in this chapter to illustrate the potential of the new modelling framework. As the pedestrian is forced to walk in a controlled manner during the lateral lock-in pedestrian test, the crowd-structure model previously described has been applied under the deterministic approach.

The natural frequency (around 0.91 Hz) and associated damping ratio (approximately 0.55%) of the first lateral vibration mode of the footbridge were identified experimentally. As the natural frequency of this vibration mode is within the range that characterizes the pedestrian-structure interaction in lateral direction, a lateral lock-in pedestrian test was conducted to determine experimentally the number of pedestrians that originates the lateral instability phenomenon [31]. The analysis focused on characterizing the beginning of the lateral lock-in phenomenon, since during this part of the phenomenon, the modification of the modal properties of the structure induced by the pedestrian-structure interaction is higher [3]. The lateral acceleration, a_{lat}, at mid-span of the structure in terms of the number of pedestrians, which cross along the structure, was recorded in this lateral lock-in pedestrian test. The analysis of the

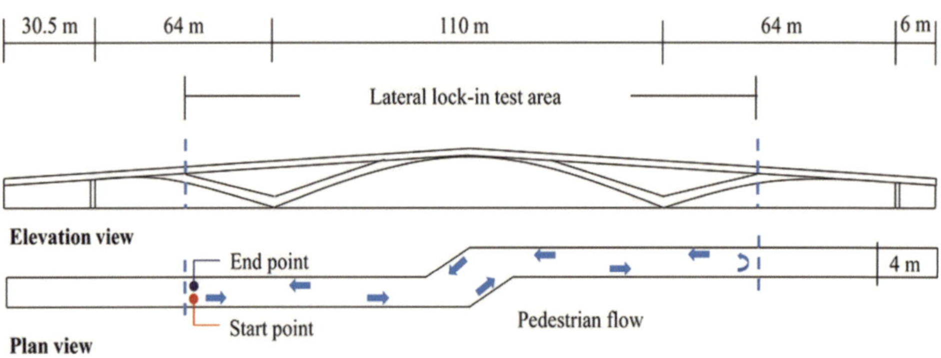

Figure 5. Scheme of the lateral lock-in pedestrian test on Pedro e Inês footbridge [31].

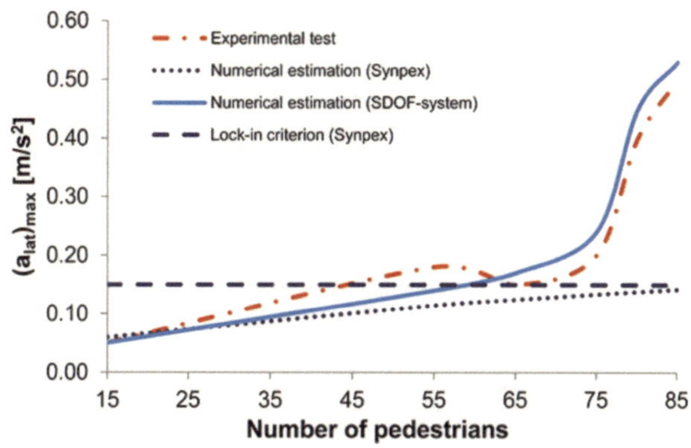

Figure 6. Experimental and numerical variation of the maximum lateral acceleration, $(a_{lat})_{\max}$, during the lateral lock-in pedestrian test [23].

graphical representation of these results (**Figure 6**) allows for identifying the beginning of the instability lateral lock-in phenomenon. As it is illustrated in **Figure 6**, the number of pedestrians that originates the beginning of the lateral lock-in phenomenon is around 75 [31].

Subsequently, a numerical lateral lock-in test based on the proposed approach was performed. Each considered group of pedestrians has been simulated considering as initial spatial distribution, a rectangular-shaped grid with an initial distance among pedestrians $d_p = 0.50$ m in longitudinal direction and an equidistant distribution in lateral direction. During the numerical test, according to the assumptions of the experimental test reported in the literature [31], each considered group of pedestrians walks freely along the footbridge, following the curve path illustrated in **Figure 5**. The number of pedestrians in each group increases gradually between 15 and 85 in increments of 5. The coordinates of the considered lateral vibration modes of the structure follow from the results available in the literature [31].

As result of this numerical analysis, the maximum lateral acceleration at mid-span of the structure in terms of the different groups of pedestrians on the footbridge was obtained. The graphical representation of this relationship is shown in **Figure 6**. A good agreement is achieved between the experimental lateral maximum accelerations and the numerically estimated maximum values, as it is illustrated in **Figure 6**. Additionally, the estimation of the numerical maximum acceleration obtained, applying the methodology proposed by the Synpex guidelines [12], is also shown in **Figure 6**. It is clear from **Figure 6** that the new modelling framework allows obtaining a more accurate numerical analysis of the lateral lock-in phenomenon than these design guidelines. The lateral lock-in criterion established by the Synpex guidelines [12] is also illustrated for reference in **Figure 6**.

8. Conclusions

The assessment of the vibration serviceability limit state of footbridges under pedestrian-induced excitation has usually been performed based on the recommendations of the most advanced international standards and design guidelines. However, the numerical estimation of the dynamic response of footbridges obtained according to these codes differs from the value recorded experimentally.

In order to overcome this problem, a new generation of crowd-structure interaction models, that constitute a new modelling framework, has been proposed by the scientific community. All these models share, as common characteristic, that they simulate the crowd-structure interaction phenomenon using two sub-models: (i) a pedestrian-structure interaction sub-model and (ii) a crowd sub-model. For the first sub-model, the pedestrian is modelled by a SDOF, MDOF or IP system and the structure via its modal parameters obtained from a finite element model. For the second sub-model, the last tendency is to use a multi-agent method based on the principles of the social force model. The linking between the two sub-models is achieved by the inclusion of several behavioural conditions in the model. Comfort and lateral lock-in threshold are usually considered. Three key aspects are taken into account for this new modelling framework: (i) the inter-and intra-subject variability, (ii) the pedestrian-structure interaction and (iii) the crowd dynamics. The last two aspects are guaranteed by the own formulation of the model, and the first is ensured assuming that the different parameters of the crowd-structure interaction model are random variables.

One of these new crowd-structure interaction models has been described briefly in this chapter, emphasizing the section corresponding to the crowd behaviour.

Finally, the potential of this new modelling framework has been illustrated with a case study, the analysis of the lateral lock-in phenomenon of the Pedro e Inês footbridge (Coimbra, Portugal). As result of this study, a good agreement is achieved between the number of pedestrians which originates the lateral instability phenomenon obtained during the experimental test and the numerical estimation determined via the crowd-structure interaction model.

9. Future trends

Although the use of the crowd-structure interaction model allows improving the estimation of the response of footbridges under pedestrian flows, further studies are being conducted in order to better characterize some aspects of these models. Among others, the following research lines may be cited:

i. The crowd-structure interaction model might be generalized to longitudinal direction via the estimation of the parameters of the SDOF-system in that direction.

ii. In order to better characterize the inter- and intra-subject variability, the statistical distributions that characterize the parameters of the pedestrian-structure interaction model should be improved via the analysis of the behaviour of other groups of pedestrians on different types of footbridges.

iii. The relationship between the parameters of the pedestrian-structure interaction model and the step frequency of the pedestrian should be further analysed.

iv. The parameters of the crowd sub-model, normally based on the results of researches of general purpose, should be estimated concretely for the case of pedestrians moving on footbridges, to improve still more the accuracy of the crowd-structure interaction model.

v. This new modelling framework allows establishing the comfort requirements directly in terms of the maximum accelerations experienced by the pedestrians (instead of the maximum accelerations reached by the deck of the footbridge). A new research line can be opened to establish more accurate thresholds which allow characterizing the vibration serviceability limit state better [14].

Acknowledgements

This work was supported by the Ministerio de Economía y Competitividad of Spain and the European Regional Development Fund under project DPI2014-53947-R.

Author details

Javier Fernando Jiménez-Alonso[1]* and Andrés Sáez[2]

*Address all correspondence to: jfjimenez@us.es

1 Department of Building Structures and Geotechnical Engineering, Universidad de Sevilla, Spain

2 Department of Continuum Mechanics and Structural Analysis, Universidad de Sevilla, Spain

References

[1] Bachmann H, Ammann W. Vibrations in Structures, Induced by Man and Machines. Structural Engineering Documents, IABSE, 1987. N° 3e

[2] Zivanovic S, Pavic A, Reynolds P. Vibration serviceability of footbridges under human-induced excitation: A literature review. Journal of Sound and Vibration. 2005;**279**(1-2):1-74. DOI: https://doi.org/10.1016/j.jsv.2004.01.019

[3] Dallard P, Fitzpatrick AJ, Le Bourva S, Low A, Smith R, Wilford M, Flint A. The London millenium footbridge. The Structural Engineer. 2001;**79**(22):17-33

[4] Dziuba P, Grillaud G, Flamand O, Sanquier S, Tétard Y. La passerelle Solférino comportement dynamique. Bulletin Ouvrages Métalliques. 2001;**1**:34-57 (in French)

[5] Jiménez-Alonso JF, Sáez A. A Controlling the Human-Induced Longitudinal Vibrations of a Nielsen-Truss Footbridge Via the Modification of its Natural Frequencies. International

Journal of Structural Stability and Dynamics. 2017;**17**(6) ID 1750061 (21 pages). DOI: http://dx.doi.org/10.1142/S0219455417500614

[6] Ricciardelli F, Demartino C. Design of footbridges against pedestrian-induced vibrations. Journal of Bridge Engineering ASCE. 2016;**21**(8). DOI: https://doi.org/10.1061/(ASCE)BE.1943-5592.0000825

[7] Fujino Y, Siringoringo D.M. A conceptual review of pedestrian-induced lateral vibration and crowd synchronization problem on footbridges. Journal of Bridge Engineering ASCE. 2016;**21**(8). DOI: http://dx.doi.org/10.1061/(ASCE)BE.1943-5592.0000822

[8] Racic V, Pavic A, Brownjohn JMW. Experimental identification and analytical modelling of human walking forces: Literature review. Journal of Sound and Vibration. 2009;**326** (2009):1-49. DOI: https://doi.org/10.1016/j.jsv.2009.04.020

[9] Ingólfsson ET, Georgakis CT, Jönsson J. Pedestrian-induced lateral vibrations of footbridges: A literature review. Engineering Structures. 2012;**45**:21-52. DOI: https://doi.org/10.1016/j.engstruct.2012.05.038

[10] Bruno L, Venuti F. Crowd-structure interaction in footbridges: Modelling, application to a real case-study and sensitivity analysis. Journal of Sound and Vibration. 2009;**323**(1-2): 475-493. DOI: https://doi.org/10.1016/j.jsv.2008.12.015

[11] SETRA/AFGC. Guide méthodologique passerelles piétonnes. Technical Guide Footbridges: Assessment of vibration behaviour of footbridge under pedestrian loading). SETRA; 2006

[12] Butz CH, Heinemeyer C, Goldack A, Keil A, Lukic M, Caetano E, Cunha A. Advanced Load Models for Synchronous Pedestrian Excitation and Optimised Design Guidelines for Steel Footbridges (SYNPEX). RFCS-Research Project RFS-CR-03019. 2007

[13] Van Nimmen K, Lombaert G, De Roeck G and Van den Broeck P. Vibration serviceability of footbridges: evaluation of the current codes of practice. Engineering Structures, 2014;**59**: 448-461. DOI: http://dx.doi.org/10.1016/j.engstruct.2013.11.006

[14] Jiménez-Alonso JF. Proposal and calibration of a biodynamic model of human-structure interaction by the resolution of the inverse dynamic problem: Application to pedestrian bridges. PhD Thesis. Universidad de Sevilla, 2015

[15] Venuti F, Racic V, Corbetta A. Modelling framework for dynamic interaction between multiple pedestrians and vertical vibrations of footbridges. Journal of Sound and Vibration. 2016;**379**:245-263. DOI: https://doi.org/10.1016/j.jsv.2016.05.047

[16] Van Nimmen K, Lombaert G, De Roeck G, Van den Broeck P. The impact of vertical human-structure interaction on the response of footbridges to pedestrian excitation. Journal of Sound and Vibration. 2017;**402**:104-121. DOI: https://doi.org/10.1016/j.jsv.2017.05.017

[17] Shahabpoor E, Pavic A, Racic V. Structural vibration serviceability: New design framework featuring human-structure interaction. Engineering Structures. 2017;**136**:295-311. DOI: https://doi.org/10.1016/j.engstruct.2017.01.030

[18] Shahabpoor E, Pavic A, Racic V. Interaction between walking humans and structures in vertical direction: A literature review. Shock and Vibration. 2016;**2016**. Article ID 34302 85:22 pages. DOI: http://dx.doi.org/10.1155/2016/3430285

[19] Helbing D, Molnár P. Social force model for pedestrian dynamics. Physical Review. 1995;**51**(5):4282-4286

[20] Carrol SP, Owen JS, Hussein MFM. Modelling crowd-bridge dynamic interaction with a discretely defined crowd. Journal of Sound and Vibration. 2012;**331**:2685-2709. DOI: https://doi.org/10.1016/j.jsv.2012.01.025

[21] Gao Y-A, Yang Q-S, Qin J-W. Bipedal crowd–structure interaction including social force effects. International Journal of Structural Stability and Dynamics 2017;**17**(7):31. 1750079. https://doi.org/10.1142/S0219455417500791

[22] Jimenez-Alonso JF, Saez A, Caetano E, Magalhães F. Vertical crowd-structure interaction model to analyze the change of the modal properties of a footbridge. Journal of Bridge Engineering ASCE. 2016;**21**(8). DOI: https://doi.org/10.1061/(ASCE)BE.1943-5592.0000828

[23] Jiménez-Alonso JF, Saéz A, Caetano E, Cunha A. A crowd-structure interaction model to analyze the lateral lock-in phenomenon on footbridges. International Journal of Computational Methods and Experimental Measurements. 2017;**6**(4):1-8 (in press)

[24] Bocian M, Macdonald JHG, Burn JF. Biomechanically inspired modelling of pedestrian-induced forces on laterally oscillating structures. Journal of Sound and Vibration Sound. 2012;**331**:3914-3929. DOI: http://dx.doi.org/10.1016/j.jsv.2012.03.023

[25] Toso MA, Gomes HM, Silva FT, Pimentel RL. Experimentally fitted biodynamic models for pedestrian–structure interaction in walking situations. Mechanical Systems and Signal Processing, 2015;**72-73**:590-606. DOI: https://doi.org/10.1016/j.ymssp.2015.10.029

[26] Caprani CC and Ahmadi E. Formulation of human-structure interaction system models for vertical vibration. Journal of Sound and Vibration, 2016;**377**:346-367. DOI: http://dx.doi.org/10.1016/j.jsv.2016.05.015

[27] Bertram JEA, Ruina A. Multiple walking speed-frequency relations are predicted by constrained optimization. Journal of Theoretical Biology, 2001;**209**(4):445-453. DOI: http://dx.doi.org/10.1006/jtbi.2001.2279

[28] Macdonald JHG. Lateral excitation of bridges by balancing pedestrians. Proceeding of the Royal Society A, First Cite, September. 2008. DOI: https://doi.org/10.1098/rspa.2008.0367

[29] Venuti F, Bruno L, Bellomo N. Crowd dynamics on a moving platform: Mathematical modelling and application to lively footbridges. Mathematical and Computer Modelling. 2007;**45**(3-4):252-269. DOI: https://doi.org/10.1016/j.mcm.2006.04.007

[30] Ronnquist A. Pedestrian Induced Lateral Vibrations on Slender Footbridges. PhD Thesis. Norwegian University of Science and Technology; 2005

[31] Caetano E, Cunha A, Magalhães F, Moutinho C. Studies for controlling human-induced vibration of the Pedro e Inês footbridge, Portugal. Part 1: Assessment of dynamic behaviour. Engineering Structures. 2010;**32**:1069-1081. DOI: https://doi.org/10.1016/j.engstruct.2009.12.034

[32] Mottershead JE, Link M, Friswell MI. The sensitivity method in finite element model updating: a tutorial. Mechanical System and Signal Processing. 2011;**25**:2275-2296. DOI: https://doi.org/10.1016/j.ymssp.2010.10.012

Chapter 6

Wind Action Phenomena Associated with Large-Span Bridges

Daniel C. Vaz, Raquel A.B. Almeida and
António R. Janeiro Borges

Additional information is available at the end of the chapter

http://dx.doi.org/10.5772/intechopen.73061

Abstract

In the past, the design of bridges over increasing distances was limited by construction techniques and, as always, by economics. As technological advances have turned possible cable-supported bridges of incredible spans, a new challenge has been added to the equation: that of withstanding the action of winds without developing undesirable dynamic responses. In this chapter, the several aerodynamic phenomena of relevance to long-span bridges are classified and discussed. This will interest both experts and non-experts in the field, thanks to the overview that is given. For certain cases, codes of practice recommend wind tunnel tests. The reader is introduced to these, as well as to numerical simulations, which are currently gaining increasing importance. Next, measures for attenuating susceptibility for undesirable dynamic responses are reviewed. The chapter ends with a discussion of the Vila Real Bridge deck section, based on wind tunnel tests and numerical simulations carried out by the authors: the aerodynamics was effectively improved with geometrically subtle modifications that were proposed and adopted still in the design phase.

Keywords: wind actions, aeroelasticity, bridge decks, wind tunnel tests, aerodynamics, attenuation measures

1. Introduction

1.1. Relevance of bridge aerodynamics

Bridges are essential to the expansion and development of societies. Having in mind that erection sites are all different, we see a wide variety of bridges and it can even be said that no bridge is equal to the other. There are types of bridges but each case is a case; bridges are not a mass-production item. Several constraints to the design are brought into play, which vary from place to place and over history, such as economics, construction means and time schedule, difficulties of construction, and architectural aesthetics. Technological advances have made possible the construction of bridges over ever-increasing spans using less mass of materials, which is of socioeconomic interest.

According to the system supporting the deck, bridges can be classified into pillar-supported, arch-supported, cable-stayed, and cable-suspended. In moving from the first to the last case in this list, and generally speaking, the main span increases and, as a result, the flexibility of the bridge. Modern materials, construction techniques, and more accurate design impart to present-day bridges high flexibility and low structural damping, which are precisely the characteristics that render them more prone to aerodynamic instabilities. Indeed, the frequencies of the natural vibration modes of such bridges are sufficiently low to be within the energetic range of frequencies of the aerodynamic phenomena and thus undesirable resonance can take place.

When the structure vibrates, as a result of the fluid motion, its movement, if of sufficient amplitude, changes the flow pattern in the immediate vicinity of the structure, which in turn changes the effect of the flow on the structure. This results in a complex and dynamic interaction, called aeroelastic interaction, that does not correspond whatsoever to the situation in which the structure remains static or quasi-static.

Bridge design should bear in mind aeroelastic phenomena, in order to keep them safe for use, that is to say, under the action of wind, bridges shall not develop dangerous or inconvenient oscillations. In this regard, the deck's cross-section geometry is of paramount importance. Since it is not imposed by aerodynamics alone, such a thing as a single optimal type of deck is not available or possible, and instead there are a number of types of decks.

To show the relevance of the study of the aerodynamics of bridge decks, we give here three examples in which aerodynamic oscillations developed and had impact in serviceability or led to modification costs.

The Rio-Niterói Bridge (Rio de Janeiro, Brazil) has been the world's longest steel-box-girder pillar-supported bridge with a central span of 300 m at 72 m high. Since it opened to traffic in 1974, low-velocity winds have often set it into oscillations [1]. When this happened, the bridge was closed to traffic, for the sake of user comfort and overall safety, though with important traffic inconveniences.

The Great Belt Suspension Bridge (Denmark) is, outside of Asia, the bridge with the longest main span: 1624 m. It opened to traffic in 1998. During the final phases of construction, low-frequency vertical oscillations of the girder were observed for lateral wind of moderate speeds (5–10 m/s) [2]. A theoretical study showed that the structural impact of the oscillations would be insignificant. However, there were concerns that the oscillations could distract drivers due to visual impact, jeopardise road safety, and convey a sense of unsafeness to the public potentially resulting in a loss in traffic volume. The oscillations were later attenuated with guide vanes mounted at the bottom edges of the girder.

More recently, in May 2010, months after being opened to traffic, the Volgograd Bridge (Russia) was closed to motor traffic for 5 days due to strong oscillations. In the autumn of 2011, this prestressed concrete girder bridge, with a main span of 160 m, was fitted with tuned mass dampers to correct the situation [3].

This chapter is aimed at readers without expert knowledge in the field of aerodynamics and the objective is to provide information that the reader can later complete by exploring in more depth the available wealth of data in the published material.

1.2. Typical bridge deck sections

While, from the functional standpoint, a bridge deck is the indispensable element of a bridge, it is also the one that is of greater concern from the point of view of the bridge's aerodynamic performance. Its shape is of paramount importance and the design has to go hand in hand with structural criteria from civil engineering and the other aforementioned criteria. Consequently, a few types of decks exist. To begin with, we can distinguish between road and railway bridges. The latter are reinforced to increase the stiffness both in torsion and in bending. Long road bridges often have wider decks in order to accommodate more lanes of traffic. In what regards wind action upon them, one can identify decks with a continuous solid surface exposed to the lateral wind, like plate-girder, box-girder, and streamlined-trapezoidal deck sections, and decks based on trusses, which are permeable to the wind (**Figure 1**).

Each type of deck exhibits distinct susceptibility to aerodynamic phenomena. Moreover, for each one, several instances of aerodynamic studies can be found in the literature for the reason that decks of very similar geometry may nonetheless exhibit considerably distinct aerodynamic performances, be it as a result of the complexity of the aerodynamic phenomena or as a result of the dissimilarity between the dynamic properties of the decks. For example, for the same overall shape, a box-girder for a cable-stayed bridge with a central plane of cables has a

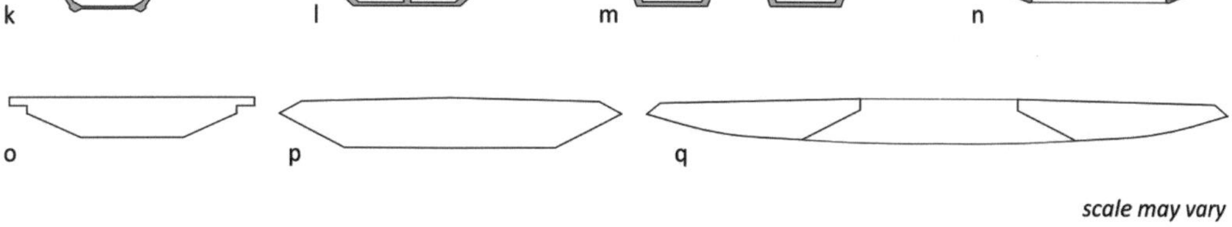

a Slab
b Rectangular
c Slab on beams
d Edge beam girders
e Rectangular box girder w/ overhangs
f Box girder w/ inclined struts
g Slab on rolled-steel girders
h Plate girders
i Truss girder (below deck)
j Truss girder (above deck)
k Trapezoidal box girder w/ overhangs
l Multiple-cell box girder
m Multiple box-girder
n Triangular edge girders
o Trapezoidal
p Streamlined trapezoidal
q Central vented dual-deck

Figure 1. A selection of types of bridge decks.

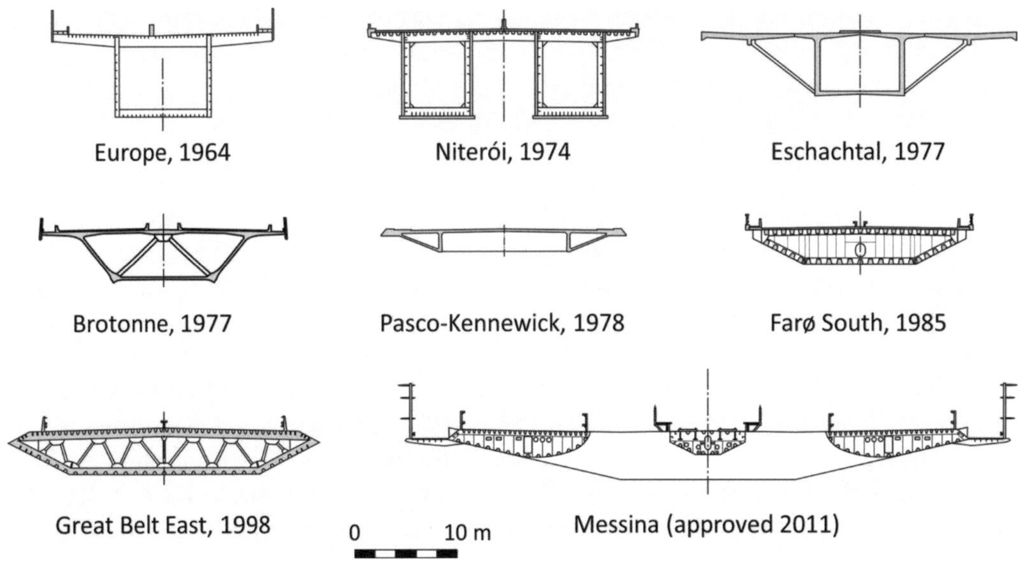

Figure 2. From general rectangular decks to trapezoidal and streamlined decks.

higher depth-to-width ratio than one for a bridge with a double plane of cables, because the girder rigidity has to compensate the lack of torsional stiffness in the first case. This difference in aspect ratio can lead to very distinct aerodynamic responses.

In box-girder decks, the quest for high aerodynamic performance has resulted in the shift from thick rectangular decks to trapezoidal and streamlined decks (**Figure 2**).

2. Aerodynamic phenomena

Wind action on bridges can adversely influence their static and dynamic stability. Along-wind deflections are examples of static reactions. These are minor problems in the sense that the existing codes of practice cover with sufficient detail current geometries of decks allowing conservative calculations. The same is not true with regards to the dynamic stability of large and very large-span bridges, which is related with vibrations or oscillations of the deck as a result of fluctuations in the pressure distribution over its surface. For very large-span bridges, a customised approach is mandatory, both analytical and experimental. The collapse of the Tacoma Narrows suspension bridge in 1940 alerted the scientific community to the importance of aerodynamic studies of wind actions on structures and to the possible occurrence of aeroelastic phenomena.

2.1. Classification

Several types of aerodynamic phenomena can develop as a consequence of interaction of actual wind with bridge decks. The turbulence of natural wind will always induce a fluctuating force

http://dx.doi.org/10.5772/intechopen.73061

on structures. This is known as buffeting and the associated displacements are usually of small amplitude, in a way that does not change the topology of the flow around the deck. In other situations, called aeroelastic phenomena, the oscillation amplitude is large enough to interfere with the airflow. In this case, the flow pattern around a deck is affected by the structure's motion itself, giving rise to a very complex interaction between the deck's motion and aerodynamic forces.

The flexibility of long-span bridges renders them highly susceptible to aeroelastic phenomena. In this context, vortex-induced vibration (VIV), torsional flutter, coupled flutter, and galloping may arise. While in VIV, the amplitude is self-limited, in the other three phenomena, the amplitude of the deck's motion tends to increase continuously and, for this reason, they are categorised as aeroelastic instabilities. If damping is insufficient, the large amplitude of the motion may cause collapse of the structure. **Figure 3** systematises the classification of flow-induced vibrations on structures.

VIV may occur at low wind speeds, and when it does occur, it is for narrow ranges of wind speed specific for a given structure (**Figure 4**). Flutter-type instabilities arise at much higher wind speeds, above a critical value. In buffeting, turbulence in the incoming flow (a large band excitation) causes a response proportional to the dynamic pressure of the wind. The next sections address in more detail each type of flow-induced vibration.

2.1.1. Buffeting

As previously mentioned, buffeting refers to structural oscillations due to the turbulent fluctuations of oncoming wind. The structure can be considered absolutely rigid since the displacements are of very small amplitude. More specifically, the amplitude is much smaller than the thickness of the boundary layer[1], and thus the vibration of the structure does not disrupt the boundary layer, thus not changing the topology of the flow around the deck.

Turbulence has a stochastic nature; pressure and velocity are random. The turbulence energy of natural wind concentrates in the lower frequency of the velocity spectrum. That is to say that if the characteristic lateral length of the structure is comparatively small (up to about 10^2 m),

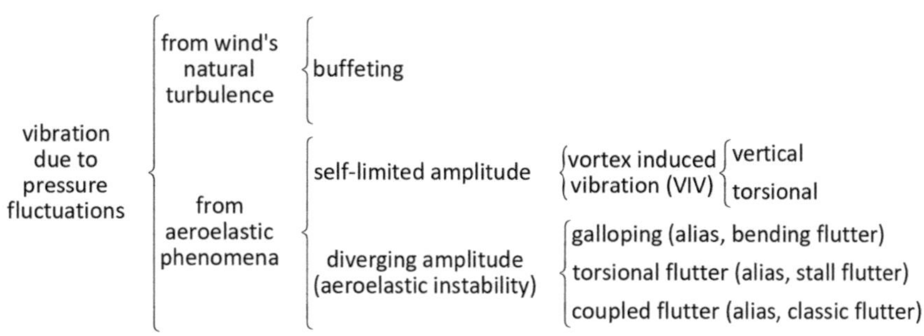

Figure 3. A classification of flow-induced vibrations, including aeroelastic phenomena.

[1]Boundary layer is the thin region near a solid surface where viscous effects are dominant and fluid velocity evolves from zero at the surface to nearly its free-stream value away from it. The depth of this layer is known as its thickness.

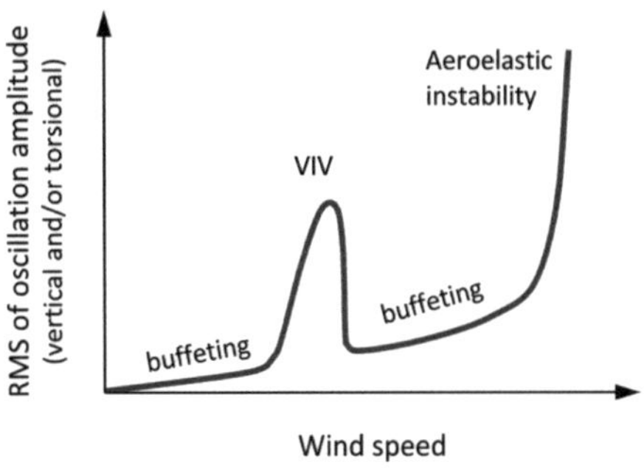

Figure 4. A generic graph of the relative amplitudes of the various flow-induced vibrations. Adapted from [4].

the aerodynamic force resulting from the turbulent flow acts as a spatially uniform fluctuating force; otherwise, the fluctuating wind can no longer be considered spatially uniform since the length scale of turbulence in the span-wise direction is comparable to, or smaller than, the span of the bridge. In this latter case, the wind peak speed effect is reduced. Indeed, the deck is not simultaneously excited over the whole of its length with the peak value of wind fluctuation; there are spatial and temporal variations.

Therefore, to optimise the dimensioning of the structure, instead of the wind peak speed, a reduced value should be used in the design. This entails the longitudinal cross-correlation of turbulent fluctuations. The problem is rather complex but the accumulated data concerning wind statistical properties and the development of analytical models associated with numerical techniques make it possible to tackle the influence of the three-dimensionality of natural wind fluctuations. From this, it is possible to derive reduction factors having in mind the relation between bridge span and turbulence length scale. If the spatial dimension of the correlation is considerably smaller than the bridge span, then the reduction factor will have a small value. Several examples of this procedure applied to bridges are available in the literature, and an interesting one concerns the Lion's Gate Bridge in Vancouver, Canada [5]. This analytical approach was developed to address the effect of gusts on aircraft wings and was transposed to civil engineering by Davenport [6].

Buffeting is proportional to the flow's dynamic pressure and thus the amplitude of buffeting-related vibration exhibits a quadratic evolution with the flow velocity. In severe cases, it can destabilise vehicles or pedestrians. On the other hand, minor vibrations, which may go unnoticed, may cause in the long-term damage by fatigue to structural components.

Even bridges with a carefully designed cross-section of the deck can still be susceptible to buffeting. Buffeting may also stem from irregular flow separations along the contour of the bridge deck, which generate a new vortex field to be added to that of the oncoming turbulence [7]. Nevertheless, if the fluctuations associated with those separations lose their stochastic nature, and periodical vortical structures are formed in the wake, then this may turn out to be what is called vortex-shedding, to be dealt with in the next section. Buffeting refers also to a different effect: increased turbulence in the wake of a structure interfering with a nearby second one placed downwind (as in the l'Iroise Bridges), and the term wake-buffeting may be used in this case (Section 6.3.3 of [8]).

2.1.2. Vortex-induced vibration

Vortex-induced vibration (VIV) can appear as either vertical or torsional oscillations. It is caused when the flow separates at spots on the deck's contour so that vortices are shed periodically from one and the other sides of the bridge deck. Flow may separate at protuberances on the upper or lower surfaces of the deck, like traffic barriers or guard-rails. A detailed discussion of the flow physics of vortex-shedding excitation can be found in two recent reviews [7, 9].

When studying the aerodynamics of the Vila-Real Bridge, the authors concluded that VIV would not occur when vortices were shed from just one side of the deck [10]. The addition of road equipment to the top of the deck raised the possibility of vortex-shedding above the deck, in addition to occurrences on the other side.

The nature of the resulting aerodynamic forces acting on the deck makes VIV self-limited in amplitude.

Vortex-induced vibrations can cause discomfort to the users of a bridge or even its failure by fatigue in the long term. It should be pointed out that the amplitude can become of concern in situations for which the system has low damping and the vortex shedding frequency is close to a natural frequency of the bridge.

Lock-in is an interesting phenomenon associated to VIV, in which the high amplitude of the oscillating motion is maintained over a somewhat wider range of wind speeds rather than for a certain wind speed.

Decks known to be particularly susceptible to VIV are plate-girder and deep box-girder decks.

2.1.3. Torsional flutter

Torsional flutter, sometimes called stall flutter, happens when the damping associated with torsional motion is very low. Here, damping is to be understood as the ensemble of structural and aerodynamic damping of the structure acted by wind. A critical velocity can be defined as the value for which the total damping becomes zero. In the onset of torsional flutter, lift causes the body to vary its pitch in a way that from its interaction with the wind a sustained torsional motion results. Once established, torsional flutter manifests as a rapid rotational oscillation.

It should be said that static deformations induced by wind loadings at velocities near flutter occurrence may possibly change the effective wind attack angle.

Torsional flutter can occur, in occasions, when propitious conditions are met, in flexible bridges with channel type or H-shaped plate girder cross-sections.

2.1.4. Galloping

Galloping, also known as bending flutter, is a single degree-of-freedom aeroelastic oscillation in the direction transverse to the wind, which can reach large amplitude. There is potential for galloping to happen whenever $C_L(\alpha)$ exhibits an accentuated negative slope. Indeed, the total damping of a bluff body undergoing wind-excited oscillations consists of a structural damping (considered positive) and an aerodynamic damping, related to $d\,C_L/d\alpha + C_D$. The system can develop divergent oscillations if the total damping becomes negative. The condition for negative aerodynamic damping to occur is known as the Den Hartog criterion [11]. Galloping occurs above this critical wind speed.

Non-streamlined box-girders with relatively small width-to-depth ratio are particularly susceptible to galloping, since the flexural stiffness is much smaller than the torsional.

Galloping can occur at low, either steady or unsteady, wind speeds, for structures with low structural damping associated with bending, which is the case of suspension bridges, although the occurrence of galloping is more typical of overhead power lines subjected to ice deposition [11].

2.1.5. Coupled flutter

Coupled flutter, also named classic flutter, encompasses motion in both bending and torsion. It occurs in cases for which the structure has a pair of neighbouring frequencies in bending and in torsion.

Bridge decks that will exhibit coupled flutter if the wind speed exceeds a critical value are those with a streamlined trapezoidal cross-section, with large width-to-depth ratio, so that there are no large vortices shedding from the leading edge. Examples are the Great Belt and the Izmit Suspension Bridges, as pointed out by [7].

Classic flutter was first perceived in aircraft wings and Theodersen [12] developed a thin airfoil theory aimed at describing analytically the flutter phenomenon. Since the aerodynamics of trapezoidal decks is far from being tractable as that of a thin airfoil, this theory cannot be directly applied to bridge decks without corrections [13]. Scanlan and Sabzevari [14] proposed a formulation in which the aeroelastic forces (lift and torque) are expressed as functions involving coefficients known as flutter derivatives. These derivatives (which cannot be obtained analytically) have to be obtained through wind tunnel testing. The resulting equations can be solved for the critical wind speed for the onset of flutter.

In certain cases, designers may resort to simpler ways of estimating this critical speed. In the beginning of the 1960s, A. Selberg proposed a formula for thin plates. Under certain conditions, it can be adapted to bridge decks through the use of a factor that depends on the geometry of the cross-section. More recently, Bartoli and Mannini [15] proposed simple expressions for critically reduced wind speed and coupling frequency based on a reduced number of flutter derivatives.

2.1.6. Comments

Current design practice avoids coupled flutter, alias classic flutter, by having in mind from the beginning of the design sufficient separation between the lowest torsional natural frequency of the bridge and the fundamental bending frequency. For example, Section E.4.2 of the Eurocode 1 [8] establishes a minimum ratio of 2:1.

In the same way, torsional flutter is not a significant phenomenon in modern bridges due to the necessary stiffness imposed by practical reasons.

Galloping can occur if Den Hartog's criterion is not observed. Vaz et al. [16] numerically studied two decks that were found stable. Nevertheless, galloping can be an important phenomenon in bridge cables, especially when the section's geometry may be changed by ice deposition.

From what has been said, it is understood that, as far as bridges are concerned, VIV is the aeroelastic phenomena that deserves the most attention. It is a major concern since the 1940s, when large oscillations began to be observed in the original Tacoma Narrows Bridge, at relatively low wind speeds.

2.2. Influence of bridge features

The occurrence of aeroelastic phenomena on long-span bridges has been widely studied. Good starting points for further reading are the reviews of Miyata [17] and of Saito and Sakata [18].

In aerodynamics, minor details (from the point of view of design) in the body contour can lead to significant modifications of the flow pattern and aerodynamic response. While this means that the aerodynamic can be improved by introducing, still at the design stage, inexpensive and simple-to-implement modifications, it also means that it is important to consider in the aerodynamic study the various stages of the deck construction since the addition of equipment, such as median dividers, guard rails, and border beams, can noticeably change the deck's aerodynamic response [17, 19–22].

2.2.1. Deck cross-section shape

The overall shape of the bridge deck's cross-section strongly influences the occurrence of aeroelastic phenomena [23–25]. A numerical study of four generic cross-section shapes developed from the well-known plate girder section of the First Tacoma Narrows Bridge, by adding horizontal plates and fairings, showed that the closed section with fairings displayed the best aerodynamic performance (in terms of drag loading and oscillating lift) and the parent H-shaped cross-section, the worst [23].

In streamlined box sections, VIV can likely be avoided or kept at insignificant levels if the angle between the horizontal extension of the deck's bottom plate and the lower inclined panels is kept smaller than 15°, in order to avoid flow separation [25]. Streamlining of the deck can possibly result in improved aerodynamic performance, with an increase in the critical wind speed for torsional flutter and decrease in the vortex-induced response [19, 26]. Even though streamlined decks have lower drag and postpone the onset of possible divergent aeroelastic instabilities to higher wind speeds, they may nevertheless exhibit oscillations at low wind speed due to vortex shedding excitation.

It should be said that trapezoidal decks are far from being tractable as thin airfoils and analytical methods, such as Theodersen's thin airfoil theory for the study of flutter, are not successful if applied without appropriate modifications. Bridge decks have to be considered as bluff bodies.

The aerodynamic stability of super-long-span bridges can be effectively improved by introducing a slot at the centre of the girder [24]. This is the design being adopted for the proposed Messina Strait (**Figure 2**) and Gibraltar Strait crossings [27, 28].

2.2.2. Railings

Changes in the porosity of the outermost railings, be it through the use of panels to deflect the wind (and thus improve the comfort of pedestrians or servicing personal) or by snow accumulation [29], would seem, at first glance, inconsequential for the bridge aerodynamics. However, making that area impermeable to the wind can render the deck section highly unstable. As the solidity ratio (considered high above 0.3) of edge safety barriers increases, so does the overall bluffness of the section, increasing drag and reducing mean lift; the effect is more relevant as the bare deck shape is more streamlined [20]. Decks with higher porosity railings have higher flutter critical wind speed [21]. The effect of the barriers contributed to the mechanism responsible for the vortex-induced oscillations of the Great Belt East Bridge in 1998 [20].

Deck equipment, such as median dividers, edge safety barriers, or parapets, can have a great impact on the bridge aerodynamics. In numerical simulations, the barriers should be included despite the increased computational effort, in order to take into account their effects on the flow [20].

2.2.3. Supporting cables

Cables are essential components of long-span bridges and they present small mass, higher flexibility in comparison to other bridge components, and low mechanical damping. Therefore, cables are even more prone to vibration than the deck.

Wind-induced flutter instability is a major concern in the design and construction of super-long-span cable-stayed bridges. While the aerodynamic contribution from the cables is generally despised when resorting to sectional models in wind tunnel experiments of cable-stayed bridges, this cannot be maintained when assessing super-long-span cable-stayed bridges. The influence of cable aerodynamic forces on the deck's flutter instability may be significant when the main span exceeds 1000 m and the frontal area (as viewed in the flow direction) of all stay cables exceeds that of the bridge deck.

3. Codes of practice

Due to the importance of the assessment of the response of bridges to the action of wind, codes of practice have been developed to aid the bridge engineer in the design calculations. They provide definitions, establish best practices, and give recommendations.

Most countries and regions of the world have their own codes of practice for bridge design, for example: "EN 1991, Eurocode 1: Actions on Structures" [8] in Europe; the "Specifications for Highway Bridges" by the Japan Road Association [30] or the "Design Standards of Superstructures for Long-Span Bridges" by the Honshu-Shikoku Bridge Authority [31], in Japan; the "LRFD Bridge Design Specifications" by the American Association of State Highway and Transportation Officials [32] in the USA; the "Wind-Resistant Design Specification for Highway Bridges" [33], in China; the "CAN/CSA-S6-00, Canadian Highway Bridge Design Code" [34], in Canada; and the Australian standard "AS 5100" [35], in Australia.

http://dx.doi.org/10.5772/intechopen.73061

In the European Union, the named Eurocodes ultimately replaced the several National Codes, with differing rules, in the various Member States. It is worth mentioning the recommendations of the European Convention for Constructional Steelwork (ECCS) [13] or, in the case of the UK up to 2010, the British standard "BS 5400" [36].

Wind actions on bridges are covered in Section 8 of the Eurocode 1 Part 1–4. It is worth pointing out that the code does not apply to bridges involving civil engineering works with heights above 200 m or bridges having spans greater than 200 m. The Vila-Real Bridge to be discussed in Section 8 exceeds both these criteria.

Codes of practice tend to be conservative though, encouraging wind tunnel tests to address the susceptibility to undesirable aerodynamic phenomena, of long-span bridges and/or with atypical shape. In their numerical studies, Bruno and Mancini [20] found for decks with complex geometries errors up to 200% between the simulated forces and the predictions using standard rules (ENV 1991-2-4). Therefore, when it is difficult to fit up a bridge within the regulation in effect, it is unavoidable to resort to aerodynamic experimental studies to obtain design criteria that warrant structural safety and, in many cases, a reduction of the overall cost of the project.

4. Wind tunnel testing

4.1. Introductory notes

It should be said that wind tunnel tests will not directly provide, as an outcome, a design of a bridge or of an optimal deck geometry non-susceptible to aerodynamic instabilities. What the design engineer can expect from wind tunnel tests is a confirmation that the design is good from the aerodynamics point of view or, otherwise, clues that will help him/her in finding the causes of oscillations and remedy them. Nowadays, it is possible, and even advisable, to complement wind tunnel tests with numerical simulations, which are addressed in the next section.

Given the discussion in the preceding sections, it is not surprising that wind tunnel testing is an integral part of the design and analysis of most long-span bridges and is often a requirement in many codes and national standards. All over the world, many laboratories equipped with wind tunnels have been committed to such studies. In these studies, the goal is to reproduce, at a reduced scale and in the best possible way, the full-scale situation in a wind tunnel and to address whether the wind action on the bridge can possibly excite any of the bridge's vibration modes. When such excitation is possible to be foreseen, then it is necessary to propose corrective modifications and test their effectiveness in the wind tunnel.

4.2. Similitude parameters

Confidence in the translation to the real prototype of the results obtained with the wind tunnel models for the dynamic behaviour of bridge deck models imposes the compliance of certain similarity criteria. In what follows, the subscripts m and p refer to the model and to the prototype, respectively.

The most natural criterion is geometric similarity, being expressed by the equation:

$$L_p/L_m = C \tag{1}$$

where C is a constant that establishes the scale factor between the model and prototype. Geometric similarity implies the homothety of the shapes of the surfaces contacting with the air, excluding of course all the details that are of very difficult construction and not significant, or even completely irrelevant given the precision with which the parameters of interest are measured.

Kinematic similarity implies that the relation between the characteristic velocities of the flow over the model, V_m, and of the oscillatory movement of the model, $f_z L_m$, is the same for the prototype. This leads to the definition of a non-dimensional parameter known as reduced velocity:

$$V_r = \frac{V}{fL} \tag{2}$$

Here, f is one of the structure's eigen-frequencies, and it is often the frequency of the fundamental vertical motion, f_z. Then, the relation between kinematically equivalent velocity scales between the prototype and the model can be defined as:

$$\frac{V_p}{V_m} = C\frac{f_{z,p}}{f_{z,m}} \tag{3}$$

The ratio of frequencies in (3) constitutes a relation for kinematically equivalent time scales, T_m/T_p, which is useful to estimate the sampling time interval in order to obtain the mean values of the parameters measured in the laboratory.

Dynamic similarity concerns the dynamic interaction between a body and the flow around it and is of a rather complex nature and thus more difficult to assure. In order to be observed, the relative magnitudes of the various forces involved in the structure's dynamics—the inertial, gravitational, aerodynamic, elastic, and structural damping forces—shall be the same for the model and the prototype, and, consequently, the motion amplitudes will be in the same proportion as the geometric scale ratio. For the airflow, the relation between inertial and viscous forces is expressed by the Reynolds number:

$$Re = VL/\upsilon \tag{4}$$

Since the values of the kinematic viscosity of air, υ, are approximately equal in the model and in the prototype, dynamic similarity would require that $V_m = C\,V_p$. This is in conflict with relation (3) established by kinematic similarity, which states that flow velocities in the wind tunnel represent much higher velocities of the flow over the prototype. However, when the model does not exhibit a streamlined section, recirculation zones are formed from flow detachments occurring at the model's sharp edges. This type of separation of the main flow from the solid surface is known to be independent of Re above a certain value. Therefore, it is important to warrant that the scale of the model and velocity in the wind tunnel result in a value of Re above that threshold.

The similarity parameter that involves the mass of the model, or of the prototype, together with the respective logarithmic decrements of the damped responses in the various oscillation modes remains to be discussed. Strictly, these variables should be addressed separately. However, when the main concern of wind tunnel studies is the analysis of vortex shedding, mass and damping can be brought together into a single non-dimensional parameter. That is to say, the system's response to the forced periodic excitation due to vortex shedding can be represented by the logarithmic damping factor, δ_s, which is itself non-dimensional. This is appropriate in situations of low damping linear systems, and in such cases, the amplitudes of the resonance peaks are inversely proportional to damping. The similarity parameter, which is named Scruton number, is defined by the following expression [8]:

$$Sc = \frac{2\,\delta_s\, m_i}{\rho\, h^2} \tag{5}$$

where m_i is a mass per unit length, ρ is the air density, and h is the cross-wind dimension (i.e., the structure's depth, specifically the section at which resonant vortex shedding occurs). Thus, the denominator in the definition represents a characteristic value of the mass of the air displaced by the structure. The Scruton number describes how sensitive a structure is to vibrating as a result of vortex shedding; the response amplitude will significantly increase at low values of this parameter.

Being extremely difficult in practice to build a model and an experimental apparatus leading to the same values of *Sc* in the model and prototype, the common practice is to build a model and suspension system as light and as least damped as possible, in order to bring out in the laboratory any tendency for instability due to the action of the flow.

4.3. Full aeroelastic bridge models vs. sectional models

A pertinent issue for researchers planning wind tunnel tests of long-span bridges is the choice between a full aeroelastic model of the bridge and a sectional model of the deck [37]. In the former, besides the full three-dimensional geometry of the bridge, the dynamic attributes of the several components must also be adequately reproduced in order to obtain a valid global response of the bridge as a whole structure and of the interaction between its structural members. For large-span bridges, full model aeroelastic studies are undertaken because of the important role of towers and cables in the overall bridge response. Nowadays, large wind tunnels exist that can accommodate this type of studies, although at the expense of smaller scales of the model.

This is not the aim of sectional model tests, where the objective is rather studying the dynamic stability of the section proposed for a bridge. Thus, a sectional model reproduces a representative section of the bridge's deck, which is extruded over a length, to produce a two-dimensional body for tests in the wind tunnel. For this purpose, the dynamic properties of the sectional model must be judiciously chosen.

Full aeroelastic model tests are usually conducted taking into consideration the effects of the terrain on the wind, whereas sectional models are usually tested in smooth flow. Therefore, in full model tests, the turbulence response is obtained directly and the effects of turbulence on flutter and vortex shedding are intrinsic and all three-dimensional aeroelastic effects are modelled implicitly. However, compared to sectional model tests, full three-dimensional tests involve higher cost and longer lead time; larger wind tunnels, which are costly to operate; and smaller scale of the model, which limits the level of detail of the flow that can be studied and possibly introduces *Re* dependency effects. Consequently, full aeroelastic model tests are often complemented with sectional model tests, as was the case in the design stage of the Trans-Tokyo Bay Bridge [38], to give an example. An interesting discussion of advantages and limitations of sectional models, taut-strip models, and full bridge models is given in [29]. Alternatively, instead of the full bridge, parts of the bridge, retaining some three-dimensional aspects, may be tested throughout the design phase; for example, at a certain point of the Messina Bridge Project, just the main span was tested at 1:250 scale.

The test of sectional models has been widely understood as a good tool to predict critical flutter instability as well as critical vortex-induced vibration. This is particularly accurate for box-girder and plate girder decks, when sectional model testing appears to predict full scale vortex shedding excitation, and there is evidence that turbulence has only but a minor effect on critical flutter speeds.

4.4. Instrumentation

To study in the wind tunnel the susceptibility of a bridge to aerodynamic instabilities, the model has to be free to oscillate in response to the action of the flow. This is a completely different situation of static testing. In what concerns sectional models, there are a few solutions, along with the associated instrumentation, to adequately support them in the wind tunnel. Typical suspension systems are based on helical springs. One configuration (used by e.g., [39]) is based on a single pair of suspension columns, which support the extremities of a shaft on the sectional model's axis of rotation. Other sets of helical springs, not contributing to the suspension though, can govern the torsional stiffness, and the setup also includes means of preventing rolling and horizontal displacements of the model. Another configuration employs four suspension columns to support the sectional model through two horizontal arms, one at each end of the model. The four columns tackle both torsional and flexural stiffness. Each suspension column includes two helical springs, one above and one below the suspension arm. Two horizontal wires limit the translation of the model in the horizontal plane. The entire assembly remains outside the test chamber. Systems of this type have been used, for example, by Larsen and Wall [40], and by the present authors [41].

Various types of sensors may be used with the suspension systems to monitor the wind-induced motions of the deck model: laser reflexion sensors, piezoelectric accelerometers, or ring strain sensors [41].

The dynamics of the flow pattern around the model can be understood using diagnostic techniques such as laser sheet visualisation (LSV) [40]. Computational fluid dynamics (CFD), to be discussed in the next section, is becoming more often used also for this purpose, and the term CFD visualisation has been employed (e.g., [10, 42]). The classic techniques of smoke lines and wool tufts are still valuable for expedite assessments.

Another possibility is the simultaneous measurement of turbulent pressure fluctuations at multiple points on the surface of the model (e.g., [43]). By analysing the spectra correlation, the development and movement of recirculation bubbles can be understood. The interest in this technique has increased at the same rate as the increase of computational power, and, nowadays, the required post-processing is done swiftly. The duration of a wind tunnel test campaign using this technique has been reduced, making them more affordable.

5. Numerical modelling

5.1. Introductory notes

At present, three-dimensional unsteady numerical simulations where all the relevant physics of the interaction of turbulent wind with a flexible structure are completely described still involve a considerable investment of time, despite the steady increase in computational power over time. The case becomes even more difficult if wind gusts, surrounding complex terrain, or coupling with neighbour bridges, are to be included. Therefore, numerical simulations are carried out to look into a certain aspect of the whole problem. Often, numerical simulations complement wind tunnel testing. The engineer has to integrate the several pieces of information to come up with a design that not only is robust from the wind action point of view but also complies with economic and structural constraints.

Generally speaking, structural aspects are studied with the finite element method (FEM), whereas aerodynamic aspects are studied with the finite volume method (FVM), or some other common method in computational fluid dynamics (CFD). It should be said that, recently, there have been advances towards using FEM for the study of the flow field.

In FEM, the structure of the bridge, or a part of it, is discretized into simple structural elements. Values of interest are then obtained at the ends (nodes) of the element, by solving the numerical model equations, and a weighting function is used within an element to obtain the values at intermediary positions on it. There is a wide variety of elements that may be used and a large choice of weighting functions ranging from simple to more elaborate and precise.

In CFD, the common approach is the Eulerian approach, in which the fluid around the deck, or pylons, or even the whole bridge, up to a long distance from the structure, is discretized into small control volumes for which transport equations (mass, momentum, turbulence, and scalars) are solved. Another approach is to employ meshless methods.

The challenge has been to couple these two tools to describe the coupling that exists in the real world between structure and wind, in order to be able to study aeroelastic phenomena.

5.2. Modelling of the structural dynamics

Finite element (FE) is the method of choice to model the structural dynamics of a bridge and can provide insight into its dynamic response, such as the main modes of oscillation and associated natural frequencies. This is especially important for bridges with low structural damping, as is the case of cable-stayed or suspension bridges. Among many studies, a few published examples on suspension bridges are The John A. Roebling Bridge in Kentucky, USA [44]; Tamar Bridge in Plymouth, UK [45]; and the Bosphorus Bridge and the Fatih Sultan Mehmet Bridge, in Istanbul, Turkey [46].

It should be noted that FE simulations are usually carried out for structural reasons, for example to assess the response to seismic or traffic loads, and therefore, its use in the assessment of the aerodynamic behaviour of the bridge [29, 47] is just another benefit to be collected from the effort put into carrying out the simulations.

Accompanying the advances in computing power, the dynamic analysis of bridges employs, nowadays, high-resolution three-dimensional FE modelling, encompassing thousands of finite elements. Among the various types of finite elements, the following should be highlighted: beam, shell and slab elements for deck and towers, and tension-only spar elements for cables. More information on the number and type of finite elements typically employed in FE analysis of bridges can be found in [46], where a number of case studies are presented.

5.3. Modelling of the aerodynamics

5.3.1. Finite volume method

Arbitrary flows of viscous fluids are described by the Navier-Stokes equations. The numerical solution of these nonlinear partial differential equations for turbulent flows is very difficult. The finite volume (FV) is the most common method used in the formulation of the numerical problem. The domain is divided into small volumes and the balance equations are solved for each, in an iterative procedure across the domain. However, the wide range of length scales imply the use of a very fine mesh up to a level that requires huge computational resources thus making direct numerical simulation (DNS) still infeasible for engineering problems. Another approach is to model the finer scales and to explicitly solve the larger turbulent scales. Yet, the resulting model, large eddy simulation (LES) [48], is still computationally intensive. For practical problems in engineering, less demanding approaches are used, in particular the one based on time-averaged equations (RANS) complemented with turbulence models. Popular turbulence models are the (κ-ε) (suitable for regions away from walls), the κ-ω (developed to tackle near wall effects), and the Reynolds stress model (which includes equations to deal with turbulent fluctuations in the three spatial directions). The single-equation Spalart-Allmaras model is also interesting for being much less computationally expensive and nevertheless useful in the study of airflow around bridge decks [16].

The problem can be further simplified by reducing the analysis to the two-dimensional space and/or by considering the flow to be steady. In two-dimensional, steady simulations, a cross-section of the deck is studied under the action of lateral wind to identify the existence of recirculation bubbles, their mean dimension, idea of the aerodynamic coefficients (C_L, C_D, and C_M), and how all of these features vary with angle of attack of the wind. Uniform wind or with shear (simulating the atmospheric boundary layer) can be studied.

In turn, unsteady simulations allow studying the pattern of recirculation bubble formation and vortex shedding around the deck's contour and their effect on the variation in time of the aerodynamic coefficients [42]. The effect of wind gusts can also be studied.

By allowing the structure's section to move in the domain in response to the aerodynamic forces acting over its contour and introducing mesh adaption techniques [49], it is possible to study fluid structure interaction, though without the three-dimensionality associated with the oscillation modes of the whole bridge.

Three-dimensional simulations are much more computationally intensive. An example of three-dimensional simulation is that of [43], in which LES has been used to obtain flutter-derivatives of a deck sectional model.

5.3.2. Discrete vortex method

Based on the existing discrete vortex method, HonoréWalther and Larsen [50] developed a relatively fast numerical model well suited for the simulation of two-dimensional bluff-body flows such as around the cross-section of bridge decks. A great advantage of this numerical tool is that it is mesh free.

The method takes the vorticity transport equation, spit into an advection part and a diffusion part, with appropriate boundary conditions at the solid surface involving the surface vortex sheet concept. A Lagrangian approach is then used in which vortex particles introduced in the domain are followed. The diffusion part is tackled via random walks. A scheme for marching in time is obtained from a Euler integration.

The input to the numerical model is the deck cross-section discretised into vortex-panels, typically a few hundreds of them. The model outputs are time series of aerodynamic force and moment, maps of pressure distribution, and streamlines. Aerodynamic derivatives can also be obtained. Larsen and Walther have applied the model to bridge decks of long-span bridges for which wind tunnel data are available and found that the model made good to excellent predictions regarding drag coefficient, Strouhal number, critical velocity for the onset of torsional and coupled flutter, and onset of vortex-induced vibrations.

5.3.3. Finite element method

While in the past FEM would not be the method of choice for solving flow equations, because unlike the FVM it would not inherently satisfy the local mass and momentum conservation, more recent advances have overcome this [51]. The new finite element procedure for the solution of incompressible Navier-Stokes equations is referred to as flow condition–based interpolation (FCBI) finite element method.

Panneer Selvam et al. [52] have used FEM together with LES to model turbulence to simulate the flow around Great Belt East Bridge approach span. The high accuracy in approximating the convection term made it possible to use a relatively modest number of nodes. Long time-steps can be used if the grid can align with the flow. The results from the 2D simulations were in reasonable agreement with wind tunnel results.

6. Attenuation of adverse aerodynamic phenomena

To begin with, it should be noted that the aerodynamic study of a deck can only be initiated after the overall geometry and main dimensions of its section have been established. Thus, at that time, considerable effort has already been put in the bridge structural design. Therefore, to avoid delays and costs, the corrective modifications to be proposed in order to diminish the possibility of wind exciting a vibration mode, as inferred from wind tunnel tests, should not lead to considerable reformulations of the initial design of the bridge. As has already been mentioned, small geometrical modifications to the bridge can lead to substantial changes of the airflow pattern around the structure and thereby modify its aerodynamic behaviour. This effect can be exploited to overcome any undesirable aspect of the bridge's aerodynamic response. Accordingly, modifications to bridge decks often consist in the addition of aerodynamic appendages, judiciously shaped, dimensioned, and positioned, or in minor changes to the original geometry.

Fujino and Siringoringo [4] review aerodynamic countermeasures that have been used in truss and box-girder bridges built all over the world to avoid aerodynamic instabilities like galloping, torsional flutter, and vortex-induced vibration. The countermeasures are specific of the aerodynamic phenomenon to mitigate.

6.1. Buffeting

The response of long-span cable-supported bridges to buffeting can be controlled through the use of auxiliary damping devices called tuned mass dampers (TMD). These are based on the secondary inertial system principle and consist of a mass attached to the structure through a spring and a dashpot. A variation of TMD is the tuned liquid damper (TLD) where the mass is replaced by a liquid (usually water).

Other types of cost-efficient dampers for mitigating buffeting response have been developed, like the tuned liquid column damper (TLCD) and its variants. This concept, first proposed by Sakai et al. [53], consists of a U-tube container in which one or more sections are partially obstructed by a plate with an orifice at the centre. The energy of structural oscillation is dissipated by the combined action of the liquid inertia, weight, and the damping effect associated with hydraulic pressure loss through the orifices within the tube. In principle, TLCD can be tuned (or retuned) to the frequency of the structure by designing it with the appropriate wetted length, or "liquid length". The advantages of TLCD systems include low cost and low maintenance, while they also represent a water reservoir that can be used for fire-fighting.

An evolution with adaptive frequency tuning capacity is the semi-active TLCD, which features an air chamber at each end of the tube where the pressures are actively controlled. Shum et al. [54] numerically investigated the performance of this system for suppressing combined lateral and torsional vibration along the construction of a real long-span cable-stayed bridge and concluded that it can effectively reduce the buffeting response of the bridge for the various construction stages. The same authors investigated a further development of the concept encompassing multiple pressurised TLCD, which provides more flexibility in the system design.

6.2. Vortex-induced vibrations

As previously mentioned, VIV is an aeroelastic phenomenon that occurs at low wind speeds and that is strongly related to the bridge section's shape. When designing the section's shape, there must be an effort towards minimising the possibility of occurrence of these phenomena [26, 40], since, generally, any ad-hoc solutions will increase the complexity of the structure and its construction and maintenance costs. If despite this effort, it is not possible to guarantee the absence of VIV, then other countermeasures have to be adopted.

One such countermeasure is the installation of tuned mass dampers (TMD), which are a secondary vibration system designed to have the same natural frequency of the main vibration system that it intends to cancel (i.e., the bridge). When the bridge oscillates, the secondary system acts as a stabiliser because of the action of its damping force on the main vibration system. However, the installation of TMD requires the existence of considerable free space in the bridge deck, which is not always available.

Over the past decades, various studies demonstrated that aerodynamic appendages, like double-flaps, flaps, or skirts, are effective countermeasures against VIV. These attachments inhibit the formation of the vortices responsible for VIV by directing high-speed flow to sweep the relevant deck's surfaces; moreover, the appendages themselves generate turbulence close to the surface of the bridge, which improves resistance to flow separation.

Other countermeasures have been proposed, like guide vanes, which have been employed for example on the Great Belt Suspension Bridge, as already mentioned in the Introduction.

6.3. Aeroelastic instabilities

Since the failure of the Original Tacoma Narrows Bridge, mainly two types of deck girders have been promoted: the streamlined box-girder and the truss stiffened girder. Wind tunnel tests came to reveal that the use of either one was not an absolute assurance of absence of flutter instabilities, and hence other countermeasures were developed.

For the truss stiffened girder deck, one countermeasure generally used is the installation of gratings on the girder, which reduces the pressure difference between the lower and the upper surfaces of the deck and thus settles flutter. This solution was adopted in the suspension bridges over the Akashi Strait in Japan and over the Tagus River in Portugal, with central spans of 1990 and 1013 m, respectively.

Flutter instabilities in plate girder decks can also be controlled with gratings. The influence of the opening ratio and the location of the grating have been object of study.

In regards to streamlined box-girders, studies have shown that the aerodynamic stability can be improved by allowing for a central slot in the box section. This increases the critical flutter wind speed of the bridge, and it further rises with the slot's width [24].

Numerous other solutions have been used to satisfy bridge construction codes' requirements for flutter. Some of them are triangular fairings [21, 26], active flaps [55], and passive flaps [56].

Galloping can particularly affect very slender box-girders. It may be difficult to control galloping through the increment of structural damping because of the small effect on the critical wind speed and given the diverging nature of the oscillation's amplitude. The countermeasure usually adopted to minimise the effects of galloping is the attachment of relatively small horizontal plates to the surface of the lower part of the girder.

7. Example of bridge design aided by wind tunnel studies: the Vila-Real Bridge

We have seen that, in what concerns to bridges, Codes of Practice strongly encourage wind tunnel tests for cases encompassing long span or great heights. If susceptibility to adverse aerodynamic phenomena is identified in the tests, then the solutions discussed in the previous section can be considered. The economics of the project favours minor geometric changes (from the structural viewpoint) comparatively to the addition of aerodynamic appendages. Therefore, wind tunnel tests and structural design should be coordinated so that any geometrical modification possibly proposed can be reflected in the final design.

In this section, the study of the cable-stayed Vila-Real Bridge, over the Corgo River Valley, Portugal, is presented as an example of the aid wind tunnel tests can provide in the design of the deck cross-section. The tests were carried out by the authors and associated team at the Laboratory for Fluid Dynamics and Applied Thermodynamics of the Department of Mechanical and Industrial Engineering (DEMI) of Universidade Nova de Lisboa (UNL).

The deck entails a concrete single-cell box-girder with the top flange extended and supported by inclined struts, also in concrete. It is a bridge cable-stayed in the midplane, and its importance among this family of bridges is made clear in **Figure 5**. Given its main span of 300 m and deck height of 230 m above the river, wind tunnel tests were due, as pointed out in Section 3 (Codes of Practice).

When the overall geometry and main dimensions of the deck section had been established, by the engineering office in charge of the bridge structural design, it was time to address, through wind tunnel tests, the susceptibility of the bridge to aeroelastic instabilities. After a small survey of aerodynamic laboratories, the study was assigned to our laboratory. Over the years, the laboratory has acquired a reasonable experience in static tests of civil engineering structures, namely viaducts, pylons of stay-bridges, telecommunication towers, and concrete poles for overhead electrical transmission lines. The purpose of the study was to examine the

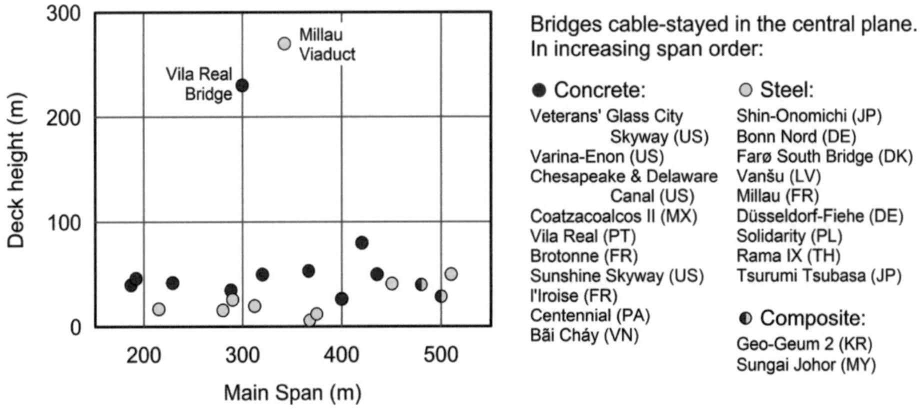

Figure 5. Deck height versus main span (adapted from [10]).

bridge's dynamic response to the action of lateral wind, for the construction and operation configurations, and to propose attenuation measures if found necessary.

Prior to wind tunnel tests, in order to understand the way the wind approaches the Corgo Valley, numerical simulations of the wind over terrain were performed with input from statistical wind data from a meteorological station a few kilometres far off.

For the reasons stated in Section 4.3, the tests were conducted with a sectional model of the deck supported in the wind tunnel by an elastic suspension system. The model, at a scale of 1:66.7, has a skeleton, of aluminium frames mounted on four steel bars, that confers the indispensable stiffness for it to not bend or twist in the dynamic tests, as well as brings the weight to the required range. The shape of the box-girder and top flange was then obtained with balsa wood lined with thin birch plywood, leaving exposed the portion of the aluminium frames that represent the inclined struts.

The suspension system, already referred to in Section 4.4 and described in detail in [41], was adjusted taking into consideration the suspended mass, the logarithmic decrement factor of the combined suspension assembly and model, and the ratio of torsional to heaving natural frequencies of the bridge that were made available by the design team from FE simulations. The adjustment was made for each configuration: bare deck during construction (with the underlying Eigen frequencies corresponding to the situation in which the bridge's central span is about to be closed) and fully equipped deck (with frequencies corresponding to a continuous construction over the whole length of the bridge).

The proper combination of forces inferred from the measurements of six ring strain sensors present in the assembly produces the dynamic action of the flow on the model in terms of lift, drag, and roll and pitch moments. The model was tested for various angles of attack and the RMS of the amplitude of oscillations both in torsion and vertical translation was obtained for the range of reduced velocities of practical interest.

Since the tests were conducted in smooth flow, buffeting shows up just slightly in the results and is possibly related to the vortex field generated by flow separations along the contour of the deck. Diverging amplitude, which would correspond to flutter, has never been found. The results for the construction configuration exhibit no peaks whatsoever, while those for the operation configuration show peaks indicating possible susceptibility to VIV. **Figure 6** shows a sample of results (angle of attack $\alpha = +2^{\circ}$). It should be pointed out that the model is more sensitive to oscillation than the real deck since, in strict terms, the sectional model represents a deck of infinite span.

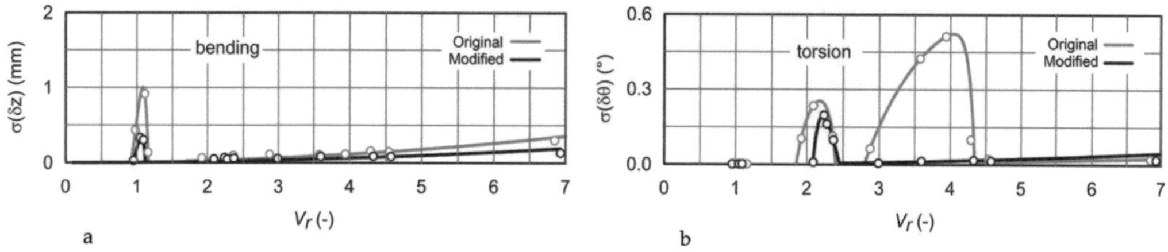

Figure 6. Results for fully equipped deck in the original and modified configurations at an angle of attack $\alpha = +2^{\circ}$. Amplitudes of: (a) vertical oscillations; (b) torsional oscillations.

To gain insight into the flow topology, two-dimensional numerical simulations of the flow were carried out, employing unsteady Reynolds-averaged Navier-Stokes (URANS) equations. By comparing comprehensively, the numerical and experimental results, it was possible to understand that, at least for this particular deck section, VIV occurs whenever the flow pattern displays simultaneously the following two features (**Figure 7**): (a) above the deck, a large coherent vortex structure on the windward part or irregular flow anywhere; (b) below the box-girder base, separated flow without reattachment. This led to the conclusion that to eliminate, or at least attenuate, the oscillatory movements of the model at certain wind speeds it would suffice to prevent vortex-shedding from at least one side of deck.

To reduce the dimensions of the recirculation bubble below the box-girder, whenever it was formed, in order to facilitate its reattachment and thus prevent vortex-shedding, the edges at the base of the box-girder were rounded. The border beam profile was also rounded. Wind tunnel tests of the deck with these slight geometric modifications showed that they were beneficial.

The geometric modifications were then proposed to the engineering office undertaking the bridge design and a meticulous discussion followed to define the transposition of the modifications to the full-scale bridge. They are visible in **Figure 8**, and the dashed lines in the sketches therein refer to the original contour.

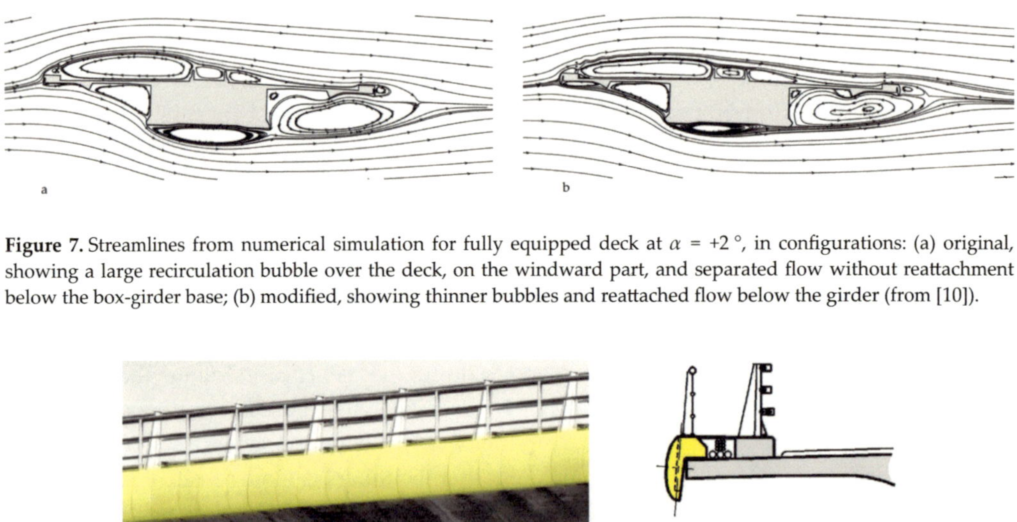

Figure 7. Streamlines from numerical simulation for fully equipped deck at α = +2 °, in configurations: (a) original, showing a large recirculation bubble over the deck, on the windward part, and separated flow without reattachment below the box-girder base; (b) modified, showing thinner bubbles and reattached flow below the girder (from [10]).

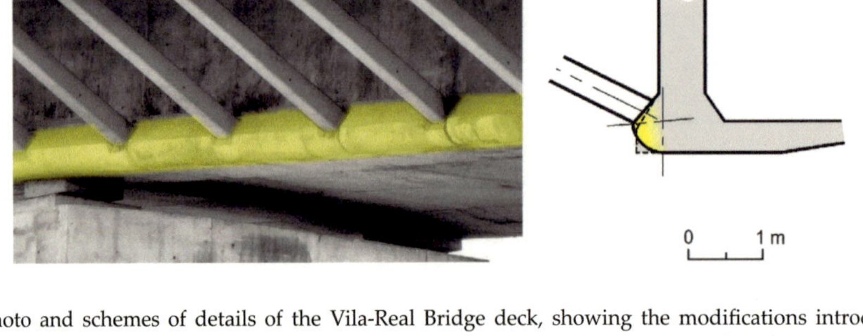

Figure 8. Photo and schemes of details of the Vila-Real Bridge deck, showing the modifications introduced by the aerodynamic study (original photo courtesy of Prof. A.P.V. Urgueira, Universidade Nova de Lisboa).

As can be seen in the results of **Figure 6**, the modifications introduced completely eliminated the prominent peak in torsional oscillations and reduced the other minor peaks (both in vertical translation and in torsion). The results in **Figure 6a** also suggest a reduction of what is one of the forms of buffeting.

8. Conclusion

The execution of projects of special structures is always complex and requires contributions from different branches of engineering. It is important that the structural design team is complemented, in due time, by aerodynamic specialists with the competence and access to the means to perform the required special studies. The Vila-Real Bridge case that has been discussed here is an example of how laboratorial means were made available, of the capacity of timely response, and of how modifications to improve the aerodynamic performance of a bridge deck can turn out to be very subtle from the geometric point of view (or even favour the aesthetics of the bridge) while being very effective with regards to aerodynamics.

It should be pointed out that major structures that very exposed to wind action are the ones that justify wind tunnel studies, and the costs of the aerodynamic studies are always a small parcel of the total investment for this kind of projects.

Acknowledgements

This work has been funded by Fundação para a Ciência e Tecnologia (PEst-OE/EME/UI0667/2014).

Author details

Daniel C. Vaz*, Raquel A.B. Almeida and António R. Janeiro Borges

*Address all correspondence to: dv@fct.unl.pt

UNIDEMI and DEMI, Universidade Nova de Lisboa, Caparica, Portugal

References

[1] Battista RC. Reduction of vortex-induced oscillations of Rio–Niterói bridge by dynamic control devices. Journal of Wind Engineering and Industrial Aerodynamics. 2000;**84**(3):273-288. DOI: 10.1016/S0167-6105(99)00108-7

[2] Larsen A, Esdahl S, Andersen JE, Vejrum T. Storebælt suspension bridge – Vortex shedding excitation and mitigation by guide vanes. Journal of Wind Engineering and Industrial Aerodynamics. 2000;**88**(2):283-296. DOI: 10.1016/S0167-6105(00)00054-4

[3] Weber F, Maślanka M. Precise stiffness and damping emulation with MR dampers and its application to semi-active tuned mass dampers of Wolgograd bridge. Smart Materials and Structures. 2014;**23**(1). DOI: 10.1088/0964-1726/23/1/015019

[4] Fujino Y, Siringoringo D. Vibration mechanisms and controls of long-span bridges. Structural Engineering International, The International Association for Bridge and Structural Engineering (IABSE). 2013;**23**(3):248-268. DOI: 10.2749/101686613X13439149156886

[5] Irwin HPA. Wind Tunnel and Analytical Investigations of the Response of Lion's Gate Bridge to a Turbulent Wind. Ottawa, Canada: National Aeronautical Establishment; 1977

[6] Davenport AG. The response of slender, line-like structures to a gusty wind. Proceedings of the Institution of Civil Engineers. 1962;**23**(3):389-408. DOI: 10.1680/iicep.1962.10876

[7] Larsen A, Larose GL. Dynamic wind effects on suspension and cable-stayed bridges. Journal of Sound and Vibration. 2015;**334**:2-28. DOI: 10.1016/j.jsv.2014.06.009

[8] EN 1991-1-4:2005. Eurocode 1: Actions on Structures - Part 1-4: General actions - Wind Actions. European Committee for Standardization; 2005

[9] Wu T, Kareem A. An overview of vortex-induced vibration (VIV) of bridge decks. Front Struct. Civil Engineering. 2012;**6**(4):335-347. DOI: 10.1007/s11709-012-0179-1

[10] Vaz DC, Almeida RAB, Didier E, Urgueira APV, Borges ARJ. Improving the aerodynamic performance of Vila-Real bridge deck-section. Journal of Wind Engineering and Industrial Aerodynamics. 2016;**156**(Supplement C):72-83. DOI: 10.1016/j.jweia.2016.07.002

[11] Den Hartog JP. Mechanical Vibrations. New York: McGraw-Hill Book Company; 1956

[12] Theodorsen T. General Theory of Aerodynamic Instability and the Mechanism of Flutter; 1934. DOI: https://ntrs.nasa.gov/search.jsp?R=19930090935

[13] Technical Committee 12 – Wind. Recommendations for Calculating the Effects of Wind on Constructions. 2nd ed. European Convention for Constructional Steelwork (ECCS); 1987

[14] Scanlan RH, Sabzevari A. Experimental aerodynamic coefficients in the analytical study of suspension bridge flutter. Journal of Mechanical Engineering Science. 1969;**11**(3):234-242. DOI: 10.1243/JMES_JOUR_1969_011_031_02

[15] Bartoli G, Mannini C. A simplified approach to bridge deck flutter. Journal of Wind Engineering and Industrial Aerodynamics. 2008;**96**(2):229-256. DOI: 10.1016/j.jweia.2007.06.001

[16] Vaz DC, Didier E, Borges ARJ. Effects of geometry modification on the aero-dynamics of a generic bridge deck section. In, JCF Pereira and A. Sequeira (Eds.): ECCOMAS CFD 2010; 2010; Lisbon

[17] Miyata T. Historical view of long-span bridge aerodynamics. Journal of Wind Engineering and Industrial Aerodynamics. 2003;**91**(12):1393-1410. DOI: 10.1016/j.jweia.2003.09.033

[18] Saito T, Sakata H. Aerodynamic stability of long-span box girder bridges and anti-vibration design considerations. Journal of Fluids and Structures. 1999;**13**(7):999-1016. DOI: 10.1006/jfls.1999.0238

[19] Bienkiewicz B. Wind-tunnel study of effects of geometry modification on aerodynamics of a cable-stayed bridge deck. Journal of Wind Engineering and Industrial Aerodynamics. 1987;**26**(3):325-339. DOI: 10.1016/0167-6105(87)90003-1

[20] Bruno L, Mancini G. Importance of deck details in bridge aerodynamics. Structural Engineering International. 2002;**5**:289-294. DOI: 10.2749/101686602777965234

[21] Wang Q, Liao H, Li M, Xian R. Wind tunnel study on aerodynamic optimization of suspension bridge deck based on flutter stability. In: The 7th Asia-Pacific Conference on Wind Engineering (APCWE-VII); Taipei, Taiwan; 2009

[22] Zhengqing C, Xugang H, Keijian O. A study of effect of cable aerodynamic forces and deck static deformation on flutter for cable-stayed bridges. In: The 5th International Symposium on Computational Wind Engineering (CWE2010); North Carolina, USA: Chapel Hill; 2010

[23] Larsen A. Advances in aeroelastic analyses of suspension and cable-stayed bridges. Journal of Wind Engineering and Industrial Aerodynamics. 1998;**74**(Supplement C): 73-90. DOI: 10.1016/S0167-6105(98)00007-5

[24] Sato H, Kusuhara S, Ogi K, Matsufuji H. Aerodynamic characteristics of super long-span bridges with slotted box girder. Journal of Wind Engineering and Industrial Aerodynamics. 2000;**88**(2):297-306. DOI: 10.1016/S0167-6105(00)00055-6

[25] Larsen A. Aerodynamic stability and vortex shedding excitation of suspension bridges. In: 4th International Conference on Advances in Wind and Structures (AWAS'08). Jeju, Korea: Techno Press; 2008

[26] Wang Q, Liao H, Li M, Ma C. Influence of aerodynamic configuration of a streamline box girder on bridge flutter and vortex-induced vibration. Journal of Modern Transportation. 2011;**19**(4):261-267. DOI: 10.1007/BF03325767

[27] COWI-"Bridge Aerodynamics", retrieved from: http://www.cowi.com/menu/service/bridgetunnelandmarinestructures/bridges/bridge-dynamics/aerodynamics/documents/021-1700-016e_09a_aerodynamics_low.pdf

[28] Diana G, Fiammenghi G, Belloli M, Rocchi D. Wind tunnel tests and numerical approach for long span bridges: The Messina bridge. Journal of Wind Engineering and Industrial Aerodynamics. 2013;**122**(Supplement C):38-49. DOI: 10.1016/j.jweia.2013.07.012

[29] Larsen A. Aerodynamic aspects of the final design of the 1624 m suspension bridge across the Great Belt. Journal of Wind Engineering and Industrial Aerodynamics. 1993;**48**(2): 261-285. DOI: 10.1016/0167-6105(93)90141-A

[30] Specifications for Highway Bridges, Japan Road Association

[31] Design Standards of Superstructures for Long-Span Bridges. Honshu-Shikoku Bridge Authority

[32] "LRFD Bridge Design Specifications" by the American Association of State Highway and Transportation Officials

[33] Wind-Resistant Design Specification for Highway Bridges (JTG/TXX-2004) (in Chinese). Beijing: People's Communication Press; 2004

[34] CAN/CSA-S6-00 (R2005) Canadian Highway Bridge Design Code

[35] Australian Standard AS 5100.2-2004, Bridge design Part 2: Design loads, section 16 - Wind loads

[36] BS 5400, National Standards Body, UK

[37] Wardlaw RL. Sectional versus full model wind tunnel testing of bridge road decks. Proceedings of the Indian Academy of Sciences Section C: Engineering Sciences. 1980;**3**(3): 177-198. DOI: 10.1007/BF02861559

[38] Fujino Y, Yoshioka Y. Wind-induced vibration and control of trans-Tokyo Bay crossing bridge. Journal of Structural Engineering, ASCE. 2002;**128**(8):1012-1025. DOI: 10.1061/(ASCE)0733-9445(2002)128:8(1012)

[39] Manzoor S, Hémon P, Amandolese X. On the aeroelastic transient behaviour of a streamlined bridge deck section in a wind tunnel. Journal of Fluids and Structures. 2011;**27**(8):1216-1227. DOI: 10.1016/j.jfluidstructs.2011.07.003

[40] Larsen A, Wall A. Shaping of bridge box girders to avoid vortex shedding response. Journal of Wind Engineering and Industrial Aerodynamics. 2012;**104**(Supplement C): 159-165. DOI: 10.1016/j.jweia.2012.04.018

[41] Almeida RAB, Vaz DC, Urgueira APV, Janeiro Borges AR. Using ring strain sensors to measure dynamic forces in wind-tunnel testing. Sensors and Actuators A: Physical. 2012;**185**(Supplement C):44-52. DOI: 10.1016/j.sna.2012.07.024

[42] Mannini C, Sbragi G, Schewe G. Analysis of self-excited forces for a box-girder bridge deck through unsteady RANS simulations. Journal of Fluids and Structures. 2016;**63** (Supplement C):57-76. DOI: 10.1016/j.jfluidstructs.2016.02.007

[43] Anina Š, Rüdiger H, Stanko B. Numerical simulations and experimental validations of force coefficients and flutter derivatives of a bridge deck. Journal of Wind Engineering and Industrial Aerodynamics. 2015;**144**(Supplement C):172-182. DOI: 10.1016/j.jweia.2015.04.017

[44] Ren W-X, Blandford G, Asce M, Harik I. Roebling suspension bridge. I: Finite-element model and free vibration response. Journal of Bridge Engineering. 2004;**9**(2):110-118. DOI: 10.1061/(ASCE)1084-0702(2004)9:2(110)

[45] Westgate RJ, Brownjohn JMW. Development of a tamar bridge finite element model. In: Dynamics of Bridges. In: Volume 5. Conference Proceedings of the Society for Experimental Mechanics Series; New York, NY: Springer; 2011. pp. 13-20. DOI: 10.1007/978-1-4419-9825-5_2

[46] Kilic SA, Raatschen HJ, Körfgen B, Apaydin NM, Astaneh-Asl A. FE model of the Fatih Sultan Mehmet suspension bridge using thin Shell finite elements. Arabian Journal for Science and Engineering. 2017;**42**(3):1103-1116. DOI: 10.1007/s13369-016-2316-y

[47] Barata V. Viaduto do Corgo da A.E. Transmontana. In: Proceedings of the Encontro Nacional Betão Estrutural (in Portuguese); Faculdade de Engenharia da Universidade do Porto-FEUP. 2012. pp. 1-20

[48] Sarwar MW, Ishihara T, Shimada K, Yamasaki Y, Ikeda T. Prediction of aerodynamic characteristics of a box girder bridge section using the LES turbulence model. Journal of Wind Engineering and Industrial Aerodynamics. 2008;**96**(10):1895-1911. DOI: 10.1016/j.jweia.2008.02.015

[49] Scotta R, Lazzari M, Stecca E, Cotela J, Rossi R. Numerical wind tunnel for aerodynamic and aeroelastic characterization of bridge deck sections. Computers and Structures. 2016;**167**(Supplement C):96-114. DOI: 10.1016/j.compstruc.2016.01.012

[50] HonoréWalther J, Larsen A. Two dimensional discrete vortex method for application to bluff body aerodynamics. Journal of Wind Engineering and Industrial Aerodynamics. 1997;**67-68**(Supplement C):183-193. DOI: 10.1016/S0167-6105(97)00072-X

[51] Kohno H, Bathe KJ. Insight into the flow-condition-based interpolation finite element approach: Solution of steady-state advection–diffusion problems. International Journal for Numerical Methods in Engineering. 2005;**63**(2):197-217. DOI: 10.1002/nme.1276

[52] Panneer Selvam R, Tarini MJ, Larsen A. Computer modelling of flow around bridges using LES and FEM. Journal of Wind Engineering and Industrial Aerodynamics. 1998;**77**(Supplement C):643-651. DOI: 10.1016/S0167-6105(98)00179-2

[53] Sakai F, Takaeda S, Tamaki T. Tuned liquid column damper-new type device for suppression of building vibration. In: Proceedings of 1st International Conference on High-Rise Buildings; Nanjing, China; 1989. pp. 926-931

[54] Shum KM, YL X, Guo WH. Buffeting response control of a long span cable-stayed bridge during construction using semi-active tuned liquid column dampers. Wind and Structures. 2006;**9**:271-296. DOI: 10.12989/was.2006.9.4.271

[55] Nissen HD, Sørensen PH, Jannerup O. Active aerodynamic stabilisation of long suspension bridges. Journal of Wind Engineering and Industrial Aerodynamics. 2004;**92**(10):829-847. DOI: 10.1016/j.jweia.2004.03.012

[56] Phan D-H, Nguyen N. Flutter and buffeting control of long-span suspension bridge by passive flaps: Experiment and numerical simulation. International Journal of Aeronautical and Space Sciences. 2013;**14**(1):46-57. DOI: 10.IJASS.2013.14.1.46

Chapter 7

Bridges Subjected to Dynamic Loading

Ján Benčat and Robert Kohár

Additional information is available at the end of the chapter

http://dx.doi.org/10.5772/intechopen.73193

Abstract

This chapter reviews the analysis of problems of highway and rail bridge dynamic response to moving traffic loads. Bridge vibrations analyses comprise solution of many interdisciplinary problems. During the two last centuries, these problems have been studied by theoretical, numerical and experimental way by many investigators. Therefore, the present chapter contains only the basic approaches for solving the complex problem of bridges subjected to dynamic loading.

Keywords: bridge structures, dynamic response of the bridge, FEM, bridge natural frequencies, moving load on bridges, full-scale bridge testing, bridges dynamic loading tests and monitoring, spectral analysis

1. Introduction

Analysis of the effects of moving loads on bridge structures was motivated by the development of rail transport in the two last centuries, which necessitated the construction of many bridges. The busy bridge transport operation eventually resulted in failures, e.g., the collapse of *Chester rail bridge* in 1947, *Takoma highway bridge* in 1940, etc. The first theoretical studies of dynamic bridge response, idealized as an elastic beam of finite length with a moving mass point, were presented in 1849 by Willis [1] and Stokes [2] and later in 1896 by Zimmerman [3]. The moving of massless force across a beam was analyzed by Krylov [4] and Timoshenko [5], who also simultaneously solved the problem of force moving across a mass-beam at constant speed. The total knowledge of the problem from that period was summarized by Inglis [6]. Nowadays, similar bridges problems are solved by numerical finite element methods via modal co-ordinate analysis of structures subjected to moving loads, e.g., Rao [7]. While the problem of rail bridge vibration has been investigated intensively since the second half of the last century, serious attempts to solve the problems of highway bridge vibration date from the middle of this century. The first report on this problem was published in 1931 by the American Society of Civil Engineers—ASCE [8] after which significant advances were made using analogue and digital computers, see also Biggs et al. [9], Looney [10], Huang and Valetsos [11], Tung et al. [12], Chaallal and Shahawi [13]. Czechoslovakia (till 1993), relevant studies were performed by Koloušek [14], Frýba [15], Baťa [16], Benčat [17], and others.

In the majority of studies, the bridge is considered as a one-dimensional beam, for which the differential equations of motion have been solved by numerical methods. The application of advanced calculation methods (finite element methods—FEM and other relevant numerical methods) enables two- and three-dimensional simulation models of bridge vibration to be solved. In the 1970s, Ting et al. [18] proposed an algorithm for solving this problem, based on an integral formulation of bridge vibration, which took into account the relations of kinematic bond of the *moving vehicle*-and-*bridge interaction*. The application of the above-mentioned methods in many cases was successful and led to the introduction of design standards and methods of assessing bridge structures. From a historical point of view, these solutions represent a gradual development in the understanding of bridge dynamic response, due to moving vehicles and their interactions.

The solution of bridge service life and reliability problems, as influenced by bridge vibration, which is mostly of random character, contributes to the complexity of this problem. However, in spite of all the complications of bridge vibration and the numerous parameters incorporated in regulations and standards in many countries, the *natural frequencies* and corresponding *modes of vibration, the dynamic coefficient* (dynamic increase of stress or deformation) and the *damping* are the basic bridge vibration characteristics, which can be verified by in situ experimental tests and monitoring.

Presently for *evaluation of dynamic response and projected parameters, new and existing bridges are utilizing numerical and experimental bridges dynamic analysis. Full-scale bridges dynamic testing and monitoring* give relevant information for projecting and assessment of real bridge behaviors [16–23]. This information consists of observed quantities obtained by experimental tests, theoretical analysis, and numerical computation and their comparison. Nowadays, the important role in the control of the bridge structures with bridge dynamic parameters (relative change of eigen-frequencies, damping parameters, fatigue parameters, vibration effective amplitudes value in time histories, etc.) plays *monitoring* of the structural parameters during normal bridge traffic on. Some results from *bridge forced vibration tests* (vibration is artificially induced (e.g., during the *dynamic loading tests—DLT*, etc.)) and also from bridge *monitoring ambient vibration tests* (input excitation is not under the control of the test engineer) are also used.

2. Theoretical and numerical approach

2.1. Simply supported beam subjected to a moving constant force

The simplest calculation model of bridge vibration is based on a simply supported elastic beam with a mass moving across the beam at constant velocity. The moving mass is assumed negligible compared with the mass of the beam. This basic case of dynamic bridge response was solved by authors, e.g., [4, 6, 13, 15] and others. The vibration caused by a force moving across an elastic Bernoulli-Euler's beam (**Figure 1**) with viscous damping, is described by the equation

$$EJ\frac{\partial^4 v(x_1 t)}{\partial x^4} + \mu\frac{\partial^2 v(x_1 t)}{\partial t^2} + 2\mu\omega_b\frac{\partial v(x_1 t)}{\partial t} = \delta(x - ct)F \tag{1}$$

with appropriate boundary and initial conditions.

The state of oscillation may be expressed as follows:

$$v(x_1 t) = v_0\sum_{j=1}^{\infty}\frac{1}{j^2\left[j^2(j^2 - \alpha^2) + 4\alpha^2\beta^2\right]}\left[j^2(j^2 - \alpha^2)\sin j\omega t - \frac{j\alpha\left[j^2(j^2 - \alpha^2) - 2\beta^2\right]}{(j^4 - \beta^2)^{\frac{1}{2}}}e^{-\omega_b t}\sin\omega'_{(j)}t - 2j\alpha\beta\left(\cos j\omega t - e^{-\omega_b t}\cos\omega'_{(j)}t\right)\right]\sin\frac{j\pi x}{l} \tag{2}$$

where

$$v_0 = \frac{Fl^3}{48EJ} \approx \frac{2F}{\mu l\omega_{(1)}^2} = \frac{2Fl^3}{\pi^4 EJ} \tag{3}$$

is the static deflection at mid-span of the beam. The *circular frequency of the damped beam vibration*, with light damping, is

$$\omega'^2_{(j)} = \omega^2_{(j)} - \omega^2_b \tag{4}$$

and with heavy damping is

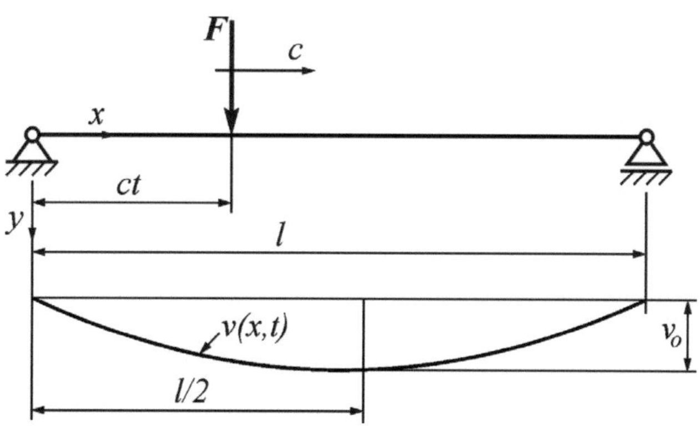

Figure 1. Simple beam subjected to a moving load.

$$\omega'^2_{(j)} = \omega_b^2 - \omega^2_{(j)} \tag{5}$$

Parameters α and β are defined as

$$\alpha = \frac{\omega}{\omega_1} = \frac{c}{2f_{(1)}l} = \frac{T_{(1)}}{2T} = \frac{cl}{\pi}\left(\frac{\mu}{EJ}\right)^{\frac{1}{2}} = \frac{c}{c_{cr}} \tag{6}$$

$$\beta = \frac{\omega_b}{\omega_{(1)}} = \frac{\omega_b l^2}{\pi^2}\left(\frac{\mu}{EJ}\right)^{\frac{1}{2}} = \frac{\vartheta}{2\pi} \tag{7}$$

The *circular frequency* of the jth mode of vibration of a simply supported beam is denoted by

$$\omega^2_{(j)} = \frac{j^4\pi^4 EJ}{l^4 \ \mu} \tag{8}$$

the corresponding *natural frequency* by

$$f_{(j)} = \frac{\omega_{(j)}}{2\pi} = \frac{j^2\pi}{2l^2}\left(\frac{EJ}{\mu}\right)^{\frac{1}{2}} \tag{9}$$

and the *circular frequency* by

$$\omega = \frac{\pi c}{l} \tag{10}$$

The solution of Eq. (2) has been analyzed by Frýba [15], with regard to parameters α and β. The maximum dynamic deflection corresponds to values $\alpha \approx 0.5$–0.7. For large values of α, deflection tends to zero, while for small values of α, deflection is practically equal to the static deflection. The critical velocities defined as

$$c_{cr} = 2f_{(1)}l = \frac{\pi}{l}\left(\frac{EJ}{\mu}\right)^{\frac{1}{2}} \tag{11}$$

are too high for practical cases. The critical velocity for the first natural frequency of steel bridges is

$$c_{cr} = 2f_{(1)}l \approx 2l\frac{10^3}{4l} = 500\,\mathrm{m/s} = 1800\ \mathrm{km/h} \tag{12}$$

The results of the theoretical analysis given in this paragraph are applicable for large-span rail and highway bridges. These bridges have very low values of the first natural frequencies and the vehicle mass is negligible compared with the bridge mass, across which they are moving. Since damping of large-span bridges is light, the dynamic displacement may be calculated from

$$v_{(x_1t)} \approx v_o \sum_{j=1}^{\infty} \sin\frac{j\pi x}{l} \times \frac{1}{j^2(j^2-\alpha^2)}\left(\sin j\omega t - \frac{\alpha}{j} e^{\omega_b t} \sin \omega_{(j)} t\right) \tag{13}$$

Eq. (13) can be simplified for low vehicle speed, $\alpha \ll 1$, into the form

$$v_{(x_1t)} \approx v_o \sum_{j=1}^{\infty} \frac{1}{j^4} \sin\frac{j\pi x}{l} \sin j\omega t \tag{14}$$

Eq. (14) represents the influence line of beam deflection at point x, expressed as a Fourier series. Since the terms for $j > 1$ are negligible, for practical applications, it is sufficient to take into account only the first term of the series

$$v_{(x_1t)} \approx v_o \sin \omega t \sin\frac{\pi x}{l} \tag{15}$$

2.2. Moving harmonic force

In the first half of twentieth century, dynamic bridges response analyses were focused mainly on studies regarding to the rail bridge vibration caused by steam traction. It has been the subject of much research (e.g., [6, 7, 14]). Inglis [7] modeled the so-called "hammer blows," due to unbalanced weights on the driving wheels of a locomotive, by a sinusoidal alternating force moving at a constant velocity across a beam. Expressing the time variation of the concentrated force by $F(t) = Q \sin\Omega t$, and considering only the first mode of beam response, the dynamic deflection in the region of resonance is given by

$$\begin{aligned} v_{(x_1t)} = v_o \frac{Q\omega_{(1)} \cos \omega_{(1)t}}{2F\ \omega^2 - \omega_b^2}\left[\omega\left(\cos \omega t - \exp^{-\omega_b t}\right)\right. \\ \left. - \omega_b \sin \omega t\right] \sin\frac{\pi x}{l} \end{aligned} \tag{16}$$

where $\omega_{(1)} = \Omega$ and v_0 is as defined by Eq. (3).

The dynamic coefficient is often defined as the *ratio of the maximum dynamic deflection to the static deflection* at the mid-span of the beam (**Figure 2**).

$$\delta = \frac{\max v(l/2, t)}{v_o} \tag{17}$$

The dependence of the dynamic coefficient on speed is sometimes called the *resonance curve*. The dynamic coefficient attains its maximum at resonance, e.g., when $\omega_{(1)} = \Omega$ and is given by

$$\delta = 1 + \frac{Q}{2F} \frac{\omega_{(1)}}{\omega^2 - \omega_b^2}\left(\omega e^{-\omega_b 1/2c} + \omega_b\right) = 1 + \frac{Q}{F}\Delta \tag{18}$$

where Δ, after substitution of the speed and damping parameters α and β from Eqs. (6) and (7) is expressed as

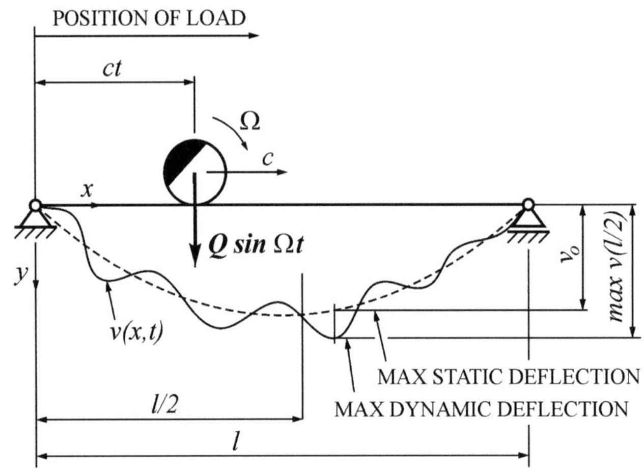

Figure 2. Mid-span deflection produced by a constant moving harmonic load.

$$\Delta = \frac{1}{2(\alpha^2 + \beta^2)}\left(\alpha e^{-\pi\beta/2\alpha} + \beta\right) \tag{19}$$

The *dynamic deformation* (or stress) *increment* may be defined as an alternative to Eq. (17), e.g., EMPA (*Swiss Federal Laboratories for Material Testing and Research*), for experimental tests of bridges, defined the *dynamic increment* ϕ [19]:

$$\phi = \frac{v_{\text{dyn}} - v_{\text{stat}}}{v_{\text{stat}}} \tag{20}$$

where v_{dyn} is the peak value of the bridge displacement measured during a passage of the test vehicle across the bridge and v_{stat} is the peak value of the bridge deflection observed under static loading caused by the same vehicle.

Application of the theoretical analysis of *bridge vibration caused by a moving harmonic force is presently not of practical significance*, due to the decline in the use of steam engines. The given knowledge, however, serves to explain the bridge vibration concepts which developed from the literature of that period. The parameter that was expressed as the *dynamic coefficient* (δ) *and its dependence on moving vehicle speed is still one of the most important parameters characterizing bridge stiffness.*

From the previous sections, it follows that *moving vehicle* on a bridge generates deflection and stresses in the bridge structure that are greater than those generated by the same vehicle applied statically. In general, the *dynamics amplification* (DA) is defined by

$$DA = \frac{R_{dyn} - R_{stat}}{R_{stat}} \tag{21}$$

where R_{dyn} and R_{stat} are maximum *dynamic and static response* (deflection, stresses, etc.) of the bridge, see also Eq. (20). Therefore, dynamic response can be calculated as

$$R_{dyn} = \text{DAF x } R_{stat} \tag{22}$$

where DAF is the *dynamic amplification factor* given by

$$\text{DAF} = 1 + \text{DA} \tag{23}$$

In addition to specifications of *most codes*, the dynamic effects of vehicles on bridges are considered by multiplying the static live loads by a *dynamic load factor* (*DLF* = δ) *greater than one*; DAF—there are many ways of interpreting this simple definition of the *DAF from test data*, see also, e.g., [19, 24].

2.3. Massless beam subjected to a moving load

The problem of a beam loaded by a moving load with negligible mass has been discussed in Section 2.1. The other extreme problem of negligible mass beam, subjected to a moving load of finite mass, was solved in [2, 3]. Consider a simply supported beam with span l and negligible mass, traversed by a load F with mass $m = F/g$, moving with a constant velocity c (**Figure 3**). Since the instantaneous position of the mass and the beam deformation are defined by a vertical displacement $v_{(a)}$ at point "a," where the force is situated, the system has one degree of freedom.

The total acting force $Y(a)$ consists of mass gravity $F = mg$ and inertia force $-md^2v(a)/dt^2$, which depends on the vertical acceleration at point $a = c\,t$, i.e.,

$$Y_{(a)} = mg - m\frac{d^2v(a)}{dt^2} \tag{24}$$

The static deflection caused by force F is given by Eq. (3) and hence the approximate solution for the dynamic coefficient δ, given by Zimmerman [3], is as follows

$$\delta = 1 + \frac{16v_o c^2}{gl^2}\left(1 + \frac{40v_o c^2}{gl^2}\right) \tag{25}$$

From Eq. (25), it is evident that magnitude of δ decreases with increasing span l. Large-span bridge structures are heavy and their mass cannot be neglected compared with the mass of moving vehicles. However, the real dynamic action of vehicles moving across short-span bridges is not reliably described by Eqs. (24) and (25). The effect of the moving mass is fairly

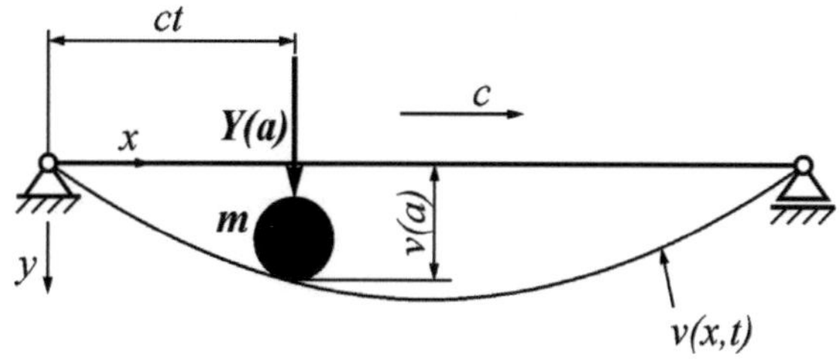

Figure 3. Massless beam with a moving mass.

small compared to that of other factors which produce high dynamic stresses in such bridges. For example, in short-span rail bridges, impact effects of flat wheels, rail joints, etc., predominate over those of the moving load. Thus, a vehicle cannot be represented adequately by a single moving point mass, even for short-span bridges.

2.4. Beam subjected to a moving system with two degrees of freedom and two axle

The need to quantify dynamic bridge response induced by moving vehicles has led to the development of improved but more complex physical models. The use of modern computers and advanced numerical methods enables satisfactory solutions of such problems to be obtained.

If the vehicle mass as well as the bridge mass is taken into account, the problem is more complicated than the problems analyzed in Sections 2.1 pending 2.3. The actual problem is described by the differential equation

$$EJ\frac{\partial^4 v(x_1 t)}{\partial x^4} + \mu\frac{\partial^2 v(x_1 t)}{\partial t^2} + 2\mu\omega_b\frac{\partial v(x_1 t)}{\partial t} = \delta(x - ct)\left[F - m\frac{d^2 v(ct,t)}{dt^2}\right] \tag{26}$$

The right-hand side of Eq. (26) expresses the motion of the force F with mass m, including the inertia effect. This problem has been discussed by [7, 9, 11–13] and others. Ting et al. [18] proposed a solution which takes into account a kinematic bond of the vehicle-bridge system. Many other solutions of this problem are published in contemporary works. These cannot be described in the context of this chapter. Therefore, only the basic formulation of the problem has been introduced to identify the parameters which influence the *bridge-vehicle system vibration*. The specific characteristics of the kinematic bond of rail bridges and highway bridges should be taken into account. A vehicle is a complex mechanical system. For the purpose of axle load calculation, it can be represented by a plane model consisting of mass points, material planes, and connecting elements. It is possible to idealize the physical model of a vehicle as a one-, two- or multi-axle system, with or without damping (**Figure 4**).

The bridge is modeled as a simply supported Bernoulli-Euler beam, with a continuously distributed mass, or as a discrete system with n-lumped masses. The surface of the beam may be assumed perfectly smooth or to have irregularities. The beam stiffness can be assumed constant or variable, based on a layered system with variable stiffness in each of the elastic layers mainly for application to rail bridges (**Figure 5**). The real behavior of the *bridge-vehicle system* can be described, more or less successfully, with the combination of physical vehicle and bridge models shown in **Figures 4** and **5**. Satisfactory results have been obtained using the vehicle models in **Figure 4(A)–(C)** for bridges with spans l > 30 m. The two- or multi-axle models of the vehicle system are more appropriate for short-span bridge investigation. The following simplifying assumptions are made in relation to the physical models:

- the load remains in contact with the surface of the bridge;
- the vehicle speed is constant;
- the bridge and vehicle damping is proportional to the velocity of vibration (viscous damping);

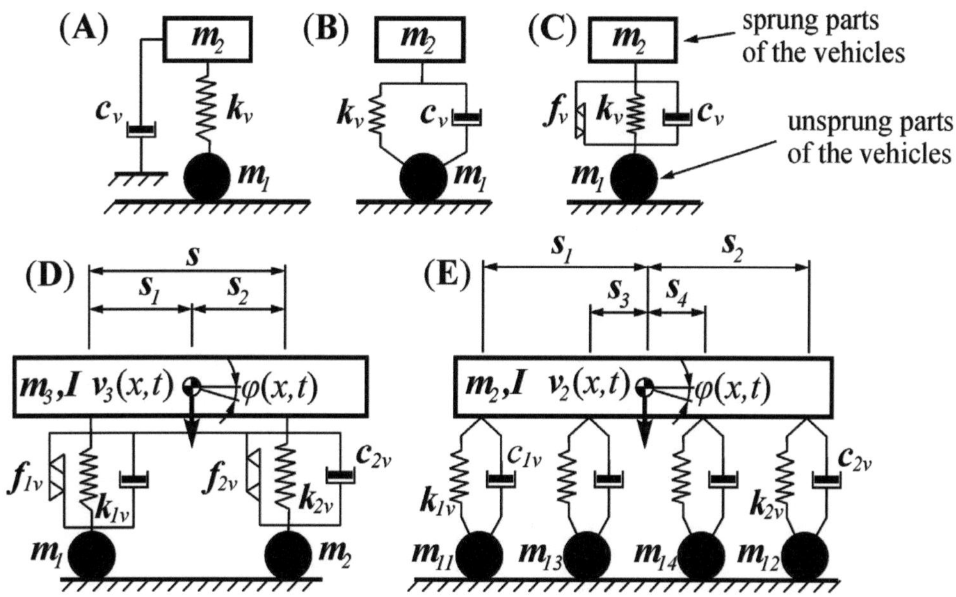

Figure 4. Physical models of a vehicle.

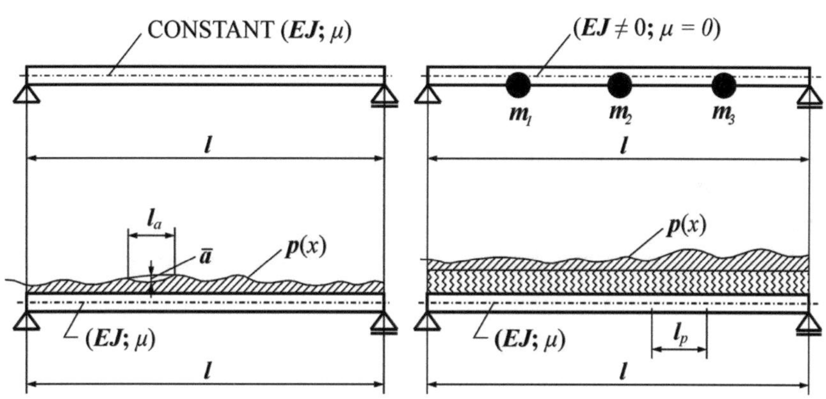

Figure 5. Physical models of bridge.

- the springing and damping of the tires are not taken into account (highway bridges);
- variable stiffness of elastic layers is taken into account for steel railway bridges (sleeper spacing effect).

The mathematical formulation of the problem of synchronous *bridge-vehicle system* vibration, taking into account the above simplifying assumptions, leads to the set of three simultaneous differential equations with variable coefficients (because of variable stiffness of elastic layers and track irregularities) describing, respectively, the vertical displacements of sprung and unsprung masses and beam vibration. The set of differential equations may be solved by numerical integration utilizing a digital computer with relevant software package.

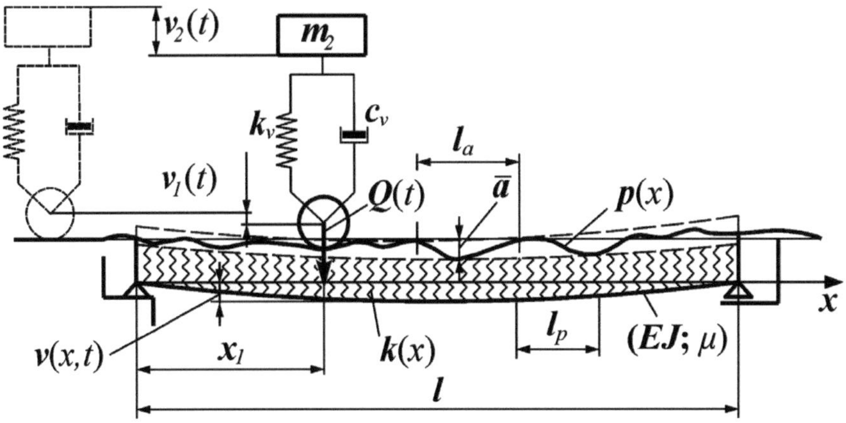

Figure 6. Model of a beam with an elastic layer and irregularities subjected to a moving system with two degrees of freedom and force $Q(t)$.

Consider the physical model of a rail bridge (**Figure 6**) with the following assumptions [15]:

1. The moving vehicle is idealized by a system with two degrees of freedom. An unsprung mass m_1 is in direct contact with the beam; m_2 denotes the sprung parts of the vehicle and the total weight of the vehicle is

$$F = F_1 + F_2 = g(m_1 + m_2) \tag{27}$$

The coordinate of the contact point is $x_1 = c.t$, because of the constant speed c along the beam.

2. The unsprung mass is acted upon only by harmonic force.
3. The top surface of the beam is covered with an elastic layer of variable stiffness $k(x)$.
4. Track irregularities are assumed to vary harmonically along the bridge span as

$$p(x) = \frac{1}{2}\bar{a}\left(1 - \cos {}^{2\pi x}/{}_{l_a}\right) \tag{28}$$

where $\bar{a}$ is the maximum depth of track unevenness and l_a is the length of track irregularity. The equations of motion of the synchronous *vehicle-bridge system* within the interval $0 \le t$ can be written

$$-m_2\frac{d^2v_2(t)}{dt^2} - k_v[v_2(t) - v_1(t)] - c_v\left[\frac{dv_2(t)}{dt} - \frac{dv_1(t)}{dt}\right] = 0 \tag{29}$$

$$\begin{aligned} F + Q(t) - m_1\frac{d^2v_1(t)}{dt^2} + k_v[v_2(t) - v_1(t)] \\ + c_v\left[\frac{dv_2(t)}{dt} - \frac{dv_1(t)}{dt}\right] - R_{(t)} = 0 \end{aligned} \tag{30}$$

http://dx.doi.org/10.5772/intechopen.73193

$$EJ\frac{\partial^4 v(x_1 t)}{\partial x^4} + \mu\frac{\partial^2 v(x_1 t)}{\partial t^2} + 2\mu\omega_b\frac{\partial v(x_1 t)}{\partial t} = \bar{\varepsilon}\delta(x - x_1)R_{(t)} \tag{31}$$

where

$$R(t) = k(x_1)\left[v_1(t) - \bar{\varepsilon}v_{(x_1,t)} - p(x_1)\right] \geq 0 \tag{32}$$

is the interactive force by which the moving system acts on a beam at the point of contact x_1, and

$$\bar{\varepsilon} = \begin{cases} 1 & \text{for} \quad 0 \leq x_1 \leq 1 \\ 0 & \text{for} \quad x_1 < 0; x_1 > 1. \end{cases}$$

Eqs. (29)–(31) should satisfy the boundary conditions of a simply supported beam as well as the appropriate initial conditions. These equations provide a very general statement of the problem of vibrations excited by a system of masses moving along a beam. Simpler sets of differential equations, which describe dynamic bridge response with sufficient accuracy, can be derived with simplifying assumptions. These equations have been solved numerically in [15].

Various individual bridge and vehicle parameters, as well as the interaction between them, were included in the theoretical analysis of the bridge dynamic response. The parameters and effects considered were: *the vehicle speed, the frequency parameter of unsprung and sprung masses, variable stiffness of elastic layers, the ratio between the weights of the vehicle and the beam, the ratio between the weights of the unsprung and sprung parts of the vehicle, the beam damping, the vehicle spring damping, the initial conditions and others.* Many contributions and solutions of this problem can be found in the literature on bridge vibrations. Wen [25] was the first author to solve this problem with application to highway bridges. In [15], the influence of the individual bridge parameters and two-axle systems on the dynamic response of steel rail bridges is also analyzed.

Rail bridges—remarks. The theoretical results have been verified by experimental tests on more than 50 rail bridges in Slovak Republic and Czech Republic and also in former Czechoslovakia (Research Rail Institute, Prague; Department of Structural Mechanics, University of Transport and Communications (UTC) Žilina; University of Žilina (1993–2017) and others).

The following conclusions can be made from the results:

1. (a) for large-span bridges with spans over 30 m, it is appropriate to consider the physical vehicle model as a moving system with two degrees of freedom (see **Figure 4(A)–(C)**); (b) for short-span bridges with spans less than 30 m, it is necessary to idealize the vehicle as a two-axle or multi-axle system.

2. The greatest influence on the dynamic increment of deflection or stress (ϕ, DA) is the vehicle speed.

3. It is necessary to include in the theoretical calculations the influence of the cross beams, uniform sleeper spacing and other regular unevenness that enlarges the local peaks in the dynamic coefficient (δ)-velocity diagram.

4. The dynamic effect of railway vehicles increases approximately in proportion to the frequency of sprung masses and the vehicle weight.

5. The dynamic stresses in short-span railway bridges are affected primarily by the impact resulting from track or wheel irregularities (rail joints, flat wheels, etc.)

6. For short-span rail bridges, the effects of sprung and unsprung vehicle masses that have been set in vibration prior to crossing the bridge are important.

7. The periodic irregularities (sleeper effects) when multi-axle vehicle systems cross the bridge can cause their vibration with resonance, especially at velocities of 100–200 km/h.

Figure 7 shows a comparison of the computed and measured deflections at mid-span of a bridge, and the dynamic coefficients δ at different locomotive speeds.

Highway bridge—remarks. The preceding discussion was directed primarily toward railway bridge vibration. Highway bridge vibration analysis should incorporate the specific features which are associated with *highway bridge structures* and *vehicle construction*, which result in different *interaction of the bridge-vehicle system*. In the case of highway bridges, the load bearing system of modern bridge structures consists mainly of prismatic and non-prismatic beams of box, open or partly closed cross section. In the majority of cases, the bridge structure approximates to the typical linear structure model. At the formulation stage of the physical model of the *bridge-vehicle system*, it is necessary to take into account the effect of variable stiffness of the roadway, which may be replaced by the effect of track irregularities. The real bridge, as well as vehicle response, can be described adequately by the physical model of the vehicle-bridge system shown in **Figure 8**. Theoretically, the problem of forced vibration of a system consisting of a moving vehicle and a bridge structure (**Figure 8**) can be described generally by operator relations, e.g. [15]

$$\begin{aligned} L_1[\{r_q(t)\}, \{r_s(t)\}, h_q(t), \{v(x_q, z_q, t)\}, a_q(t)] = 0 \\ L_2[\{v(x, z, t)\}, \{r_q(t)\}, h_q(t), a_q(t), \{v(x_q, z_q, t)\}] = 0 \end{aligned} \tag{33}$$

where L_1 and L_2 are linear or non-linear operators; $\{r_q(t)\}$ and $\{r_s(t)\}$ are displacement vectors of vehicle elements, conditioned by upper and lower links; $\{v(x, z, t)\}$ is a vector of bridge

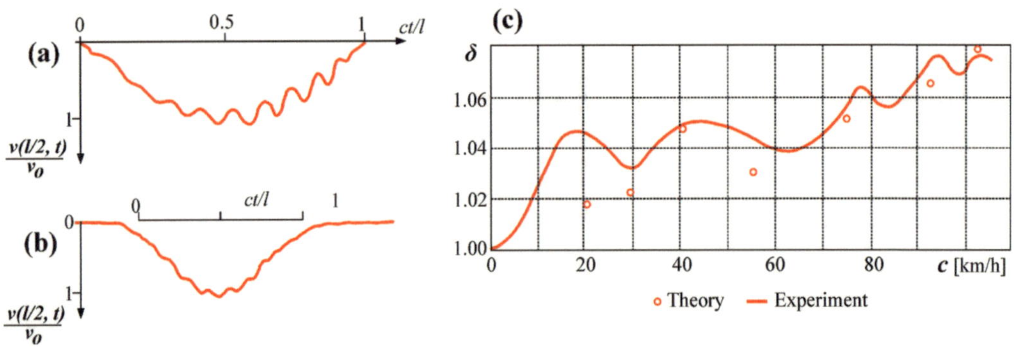

Figure 7. Deflection at the center of beam with span l = 34.8 m, traversed by an electric locomotive E 469 at speed c = 40.7 km/h [15]: (a) theory, (b) experiment (c) theoretical and experimental dependence of the dynamic coefficient δ.

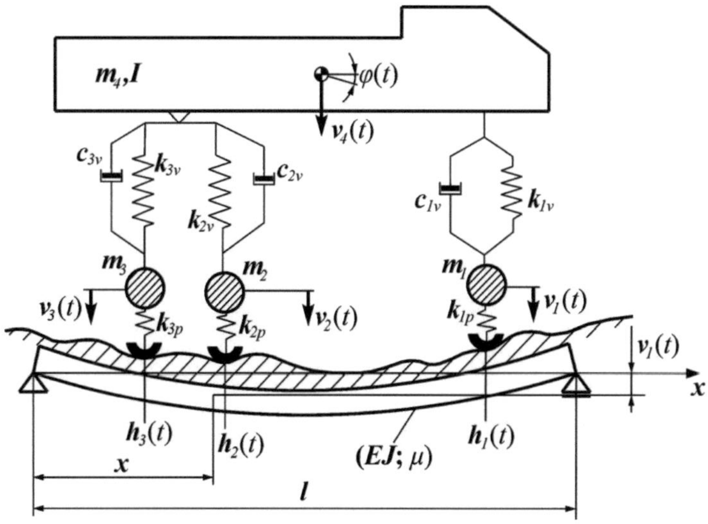

Figure 8. Physical model of a system consisting of a bridge structure and a moving vehicle.

structure displacements at the points with coordinates x, z at time t; $\{v(x_{q,}\ z_q,\ t)\}$ is a vector of bridge structure displacements at the qth lower link between the vehicle and the bridge structure; $a_q(t)$ is the law defining the vehicle movement along the longitudinal bridge axis; $h_q(t)$ is the function describing the irregularities of the road surface; and x_q, z_q are the coordinates of the lower link between the vehicle and the bridge structure. Eq. (33) by the differential operators, together with boundary and initial conditions, defines the motion of a system consisting of a moving vehicle and a bridge structure.

Research into highway bridges has not been systematic, either in Europe or elsewhere, and in some countries, it has been limited mainly to random tests of extraordinary bridge structures, before being put into operation. This has led to a variety of methods for calculating dynamic effects in individual countries, particularly in the provisions concerning dynamic coefficients (δ, DA, DLA) in respective standards or bridge regulations. At the present time, much information is available on the forced vibration of highway bridges, which should be taken into account during the formulation of the vehicle-bridge physical model. The results of theoretical analysis and parametric studies of highway bridge response, as well as the results of experimental bridge investigations, performed by the relevant research divisions of the UTC Žilina or University of Žilina (UZ Žilina) on more than 60 highway and road bridges, are now discussed in detail. The dynamic bridge response is influenced primarily by:

1. As for railway bridges, vehicle speed has the greatest influence on the dynamic increments of stress and deflection of highway bridges.

2. The first mode natural frequency of vibration of the bridge in the vertical plane (bending) and the natural frequency of vibration of the sprung vehicle mass in the vertical direction. The theoretical dynamic coefficient δ (DLA) and also experimental δ_{obs} (DAF) are maximum when $\omega_{(1)} \approx \omega_v$. It was noted also that the influence of frequency ratio $\omega_{(1)}/\omega_v$ diminishes with increasing mass ratio (m_v/m_{br}).

3. The vehicle vibration at the moment the vehicle enters the bridge, since the vehicle's energy of vibration is the primary source of the dynamic bridge response. The vertical amplitude of the vehicle vibration is decisive. The initial angular amplitude of the vehicle's sprung mass vibration can be neglected in the analysis.
4. The character of the road irregularities (joints, potholes, inserted hinges, frozen snow, etc.).

The effects of the damping of the vehicles and bridges, as well as the ratio of the sprung vehicle mass to the bridge mass, are not significant for *long-span bridges*. However, they are a significant influence on *short-span bridges vibration*. It was confirmed by the theoretical analysis and the experimental tests that the curve expressing the dependence of the dynamic coefficients on the vehicle speed is not a smooth curve but has many local projections and branching points [26].

3. Natural frequencies and modes of bridge vibration

Calculation of the natural frequencies and corresponding modes of vibration forms a basis for the determination of the dynamic characteristics of bridge vibration. At present, bridges, especially of larger spans, are complicated space structural systems, comprising many elements which interact with one another, e.g., continuous beams, framed structures, arch construction, suspended structures, and others. Principally, three different simulation models of the bridge can be used: the *discrete model*, the *model with continuously distributed mass* and *models formulated by the* FEM. Advanced numerical methods FEM have been widely developed for practical application in this field, and they are described in the technical literature and available as computer software. Therefore, only two calculation methods are shortly described in this section. The theoretical determination of the natural frequencies and modes of vibration of such structures is fairly difficult in most cases and their verification is advised by experimental measurements.

3.1. Multi-degree of freedom systems

The equations of motion of a multi-degree of freedom system take the form

$$\mathbf{M}\ddot{\mathbf{u}} + \mathbf{C}\dot{\mathbf{u}} + \mathbf{K}\mathbf{u} = \mathbf{f} \tag{34}$$

where $\mathbf{f}$ is a column matrix of applied forces and $\mathbf{u}$ is a column matrix of displacement components. Both $\mathbf{f}$ and $\mathbf{u}$ correspond to the same set of points on the structure and the same directions at these points. $\mathbf{M}$, $\mathbf{C}$, and $\mathbf{K}$ are the inertia, damping, and stiffness matrices corresponding to the displacement components $\mathbf{u}$. If the applied forces and damping forces are absent, Eq. (34) becomes

$$[\mathbf{K} - \omega^2\mathbf{M}]\{\boldsymbol{\phi}\} = 0 \tag{35}$$

if the motion is assumed to be harmonic, that is, $\mathbf{u} = \boldsymbol{\varphi}\,\exp\{i\omega t\}$. This equation is *a linear eigen problem* similar to *n*-DOF system. The eigenvalues, ω_i^2, represent the squares of the natural frequencies and the eigenvectors, ϕ_i, represent the shapes of the corresponding modes of free

http://dx.doi.org/10.5772/intechopen.73193

vibration. An eigenvector is arbitrary to the extent that a scalar multiple of it is also a solution of Eq. (35). It is convenient to choose this multiplier in such a way that ϕ has some desirable property. Such *eigen vectors* are called *normalized eigenvectors*.

3.2. Numerical procedure application in bridge structure dynamic analysis

To avoid creating complicated and sophisticated numerical models involving extensive assumptions in modeling (boundary and initial conditions, mechanisms of bridge flexibility and energy dissipation, inertia, etc.), it is useful to develop an appropriate model with realistic prediction of their dynamic response upon the comparison of the experimental results and theoretical predictions. This enables also the realistic and optimal economical designs. Nowadays, very popular and useful numerical method for engineering analysis is finite element method. FEM is a numerical procedure for obtaining solutions to many of the problems encountered in civil and structure engineering. Numerical solutions of the bridge dynamic analysis problems in many cases need experimental verification in situ, e.g., [20, 27, 29, 30, 31].

To create relevant analytical models with real dynamic bridge structure with input parameters, it is useful to apply experimental modal analysis (EMA) which provides mainly *structure natural modes engine frequencies and damping parameters* of the tested bridge structure [31]. For such type of bridge dynamic tests performance in most cases, the real bridge service conditions are too restrictive for performance such type bridge tests. In these cases, operational modal analysis (OMA) procedure is applicable, which enables to perform bridge dynamic testing and also bridge health monitoring measurements without interrupting bridge service. A well-presented review of bridge testing methods explaining their conditions, advantages and limitations was presented by Salawu and Williams [27].

The bridge dynamic analysis programs are commonly available and computational problems are not complicated to solve. A lot of FEM software packages are used in this field mainly for structures modal analysis and dynamic response of bridges (ANSYS Civil FEM Bridge, BRASS, BRIDGES, BridgeSoft, BRIDGADES (ABAQUS), ADINA, DYNSOLV, LUSAS, etc.).

4. Dynamic loading tests of bridges and monitoring

In situ dynamic testing of bridges gives very useful information for numerical modeling and assessment of real bridge dynamic parameters and service conditions. In many countries, the requirement of putting the bridges into operation is the execution of *bridge static and dynamic loading tests*, which aim is *to prove* and *confirm* the projected parameters (standards criteria, serviceability, safety limit states, etc.) of tested bridge structures according to technical standards, e.g., in Slovakia by standard STN—Slovak Technical Standards [28]. Results from *static or dynamic test enable to calibrate a bridge analytical model and can be utilized as basic data for a bridge health monitoring program and for other sophisticated calculations of the bridge dynamic response* (seismic, fatigue, etc.).

4.1. Test procedures

In this section, bridges dynamic test procedure is shortly described. Bridges are tested according to the rules of the *dynamic loading test* (DLT) [28]. Excitation of *highway bridges* are commonly due to the passage of single, fully loaded, *multi-axles lorries*. The testing vehicles' gross weight usually lies near the legal limit which is defined by standards and regulations. In the case of *railway bridges*, locomotives are used. Also *normal traffic flow* is used for both highway and railway bridges.

For the expected *dynamic bridge response* caused by well-defined individual testing vehicles, *dynamic calculations* are carried out before the bridge dynamic tests. The testing vehicle is driven with a constant speed (in each measurement travel) along the bridge and respectively in the same direction or in both directions. The tests begin with a vehicle speed of $c = 5$ km/h, which is increased after each passage in steps of 5 km/h, up to the maximum achievable speed [22, 26]. If a *static test* with the used testing vehicle is not performed before the *dynamic tests*, the bridge deflection caused by vehicle traveling at a speed $c = 5$ km/h can be considered to present the static deflection w_s with sufficient accuracy (e.g., via filtering signals).

In the case of *highway bridges*, the tests on the undisturbed bridges pavement are also repeated with a plank or *standard obstacles* placed across highway pavement, **Figure 9(a)**. The cross section of the standard obstacle (length 5000 mm) is a cylindrical sector of height 60 mm and chord length 500 mm [28]. During the tests of highway bridges pay load, tires and tire pressure are kept the same; it means that the vehicle dynamic properties remain approximately constant.

Pulse forces produced by the ignition of *pulse rocket engines* (PRE) during DLT are also used mainly on large bridge structures. *Harmonically variable forces* produced by *vibration exciters* and the free vibrations of the bridge are also applied.

This type of the DLT so-called proof-loading test is performed for checking if the construction of the bridge has been constructed according to the design project. These tests (DLT) comprise the evaluation of the *dynamic loading allowance* (DLA) → from standard = dynamic coefficients

Figure 9. Testing load: (a) lorries traveling over the obstacles; (b) the PRE applied mainly *on large bridge structure.*

δ calculated by designer and *dynamic amplification factor* (DAF—from DLT), greater than 1 (the amount by which the static effects are increased by bridge-vehicle interaction contribution).

Note 1 In relevant standards of many countries, it is defined how to obtain coefficient DLA (= δ) in a normalized way (e.g., Canada, France, Germany, India, Spain, Switzerland-UK, USA, and former Czechoslovakia, etc.).

Note 2 Nowadays for Slovak Republic (CEN member), from 1.5.2006, it is mandatory to applying new European Standard. For the chapter content, it is actual Eurocode 1—Action on Structures, Part 2: Traffic loads on bridges (EN 1991–2).

Note 3 In this EN, for road bridges, the dynamic amplification was included into the *load models* (fatigue accepted), although established for a medium pavement quality and pneumatic vehicle suspension, which depends on various parameters and on the action effect under consideration. Therefore, *it cannot be represented by a unique factor.* For example, in former Slovak Standard (till 2006)—STN 73 6203, *Load actions on bridges* for calculation of δ = DLA was used a formula in unnumbered format

$$\delta = \frac{1}{0.95 - (1.4l)^{-0.6}}$$

Note 4 For railway bridges, the dynamic amplification was accepted and dynamic factor Φ (= DLA = δ) is possible to calculate according to the given algorithm (EN 1991–2, Section 6).

4.2. DLT data acquisition and recording

In this section, *data acquisition and recording* (DAR) processes during a bridge DLT are shortly described. More detailed DAR processes descriptions are in [23, 26, 29]. Dynamic deflections are measured by *pick-ups* at the characteristic points of the bridge, which is normally at the mid-span. The bridge structure dynamic response at these points, in the horizontal and vertical directions, is then recorded in the form of time histories signal. Deflections $w(t)$ are also measured at additional points along the super-structure. Except for dynamic deflections, other relevant parameters are measured: speed of the loading vehicles, magnitude and time history of excitation forces, temperature of the structure and ambient air, wind velocity, etc.

Instrumentation: During standard dynamic tests of bridges (**Figure 10(a)**), *inductive displacement transducers*—IDT are mounted at the bridge parapet or bottom of the bridge structure, which are used to monitor displacement amplitudes time histories. In these cases, recorded displacement amplitudes time histories contains both static and dynamic components of the bridge dynamic response. The measured baseline is given by an invar wire (max 30 m), strained between the measuring points of the structure and a fixed reference point under the bridge structure. The application IDT enables extracting the static component from displacement time histories $w(t)$ by using filtering techniques. This procedure is applied for DAF calculation. When the measured structure cross section is situated over water (e.g., river, lake, bay, etc.), the IDT are usually replaced by *accelerometers* or *velocity-type transducers*, **Figure 10(b)**, or *strain gauges*, **Figure 10(c)**, (measuring of strain amplitudes time history contains both static and dynamic components of the bridge response due to moving load) with relevant hardware

Figure 10. Examples of bridge DLT instrumentation: (a) inductive displacement transducers (IDT); (b) accelerometers set up; (c) strain sensor (d) charge amplifiers devices; (e) accelerometer installation process to bridge bottom for DLT.

components (amplifiers, cables, wireless technique, etc.), **Figure 10(d)**. The view of accelerometers installation on bridge bottom for DLT is showed in **Figure 10(e)**.

The signals from the used pick-ups are amplified and filtered by the signal amplifiers and low-band pass filters, and then recorded by portable notebook with relevant software and hardware facilities in test *measuring station* (MS). Scheme and view of the equipment in MS used in situ tests are plotted in **Figure 11**. During DLT, the signals transmission from measurement

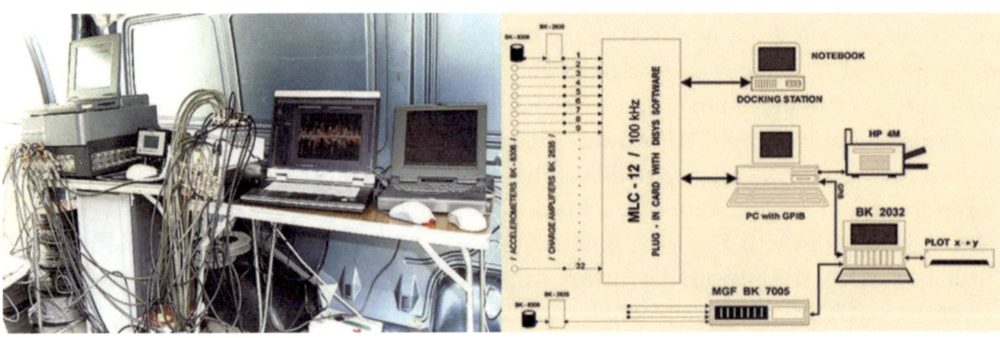

Figure 11. View of the equipment in MS with its scheme mounting used during the bridge dynamic tests.

http://dx.doi.org/10.5772/intechopen.73193

devices to the recording technique at the MS by special low noise cables were usually used. The application of the *wireless sensor network* (WSN) platform for DLT or distributed measurement application, e.g., bridge structural health monitoring, is possible to eliminate the need for costly and work-intensive wiring measuring technique. The WSN platform simplifies remote monitoring applications and delivers low-power, reliable measurement nodes that feature local control capabilities.

The application of the *quick-setup WSN* enables to implement a stand-alone remote monitoring system or easily connect with measuring PC and control systems (e.g., NI WSN, BK PULSE WSN). **Figure 12** shows examples of equipment set for experimental measurement data wireless transmissions by NI WSN modules with portable PC layout and scheme.

The final experimental analysis is usually carried out in the laboratory. The bridge vibrations induced by the lorries crossing the bridge during the DLT with different velocities are analyzed in order to quantify and compare the different dynamic effects on the bridge structure. The analysis of the time histories of vibration recorded, when lorries crossed the viaduct bridge, is processed with the following operations:

- double integration of the accelerations to displacements and evaluation of their maxima and RMS values. The displacements maximum value is used for DAF calculations;
- offset and linear trends removal;
- digital filtering with a low-pass Butterworth filter with a cut frequency, e.g., of 150 Hz and with a high-pass Butterworth filter with a cut frequency of 0.5 Hz; and
- maximum and RMS values of acceleration amplitudes evaluation.

Also from the bridge dynamic response and free vibration measured time histories, can be obtained:

- frequencies of one or more vibration modes of the loaded and unloaded bridge;
- the natural vibration damping parameters, dominant in free decay;

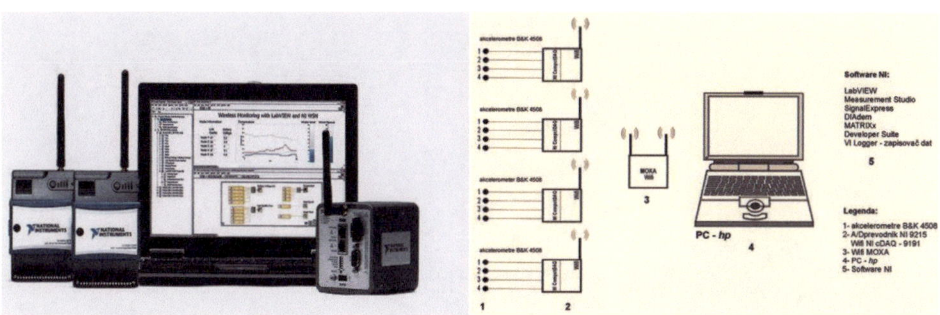

Figure 12. The experimental measurement data wireless transmissions by NI WSN modules with portable PC layout and scheme.

- percentage of vibration critical damping via the 3 dB band with method and curve fitting techniques;
- DAF and its dependence on vehicle speed;
- dynamic and static deflections as well as stresses magnitude in certain important parts of the bridge structure; and
- time history of vibration amplitudes and their classification by classification method (e.g., rain-flow classification method).

4.3. Bridges' dynamic parameters monitoring

Long-term bridges observation is discussed in literature, e.g., [29–37]. Some dynamic methods used by other authors [33], were applied to correlate relative changes of material, frequencies, and damping with carrying capacity. It was found that used monitoring techniques gave an early indication of incipient deterioration. The main scope of monitoring tests was to evaluate mainly the *relative change of well-defined natural frequencies or the corresponding damping and the RMS value of the displacements amplitude* of the bridge vibration due to traffic loading. The monitoring technique based on measurement of the bridge vibration time history due to regular traffic is not focused to give detailed bridge information but for making decision if more detailed bridge assessment methods should be used. The sophisticated bridge monitoring was introduced e.g. on the *Akhashi Kaikyo* bridge in Japan, (**Figure 13**) completed 1998. At that time, it was the largest and longest suspension bridge in the world. Bridge is a 3-span 2-hinged bridge with steel-truss-stiffened girders located near Kobe City.

Bridge has a impressive 1991-m center span between two main towers that rise 300.0 m above the sea level. The *Akashi Kaikyo* bridge, being easily affected by natural conditions and traffic means, requires high level of disaster prevention and bridge structure functionality with projected structure parameters. Therefore, to provide centralized control, traffic control and bridge structure and facility monitoring have been integrated into the Traffic Control Center. *There, information acquisition and processing are performed continuously 24 h a day, providing vital traffic and bridge structure information.*

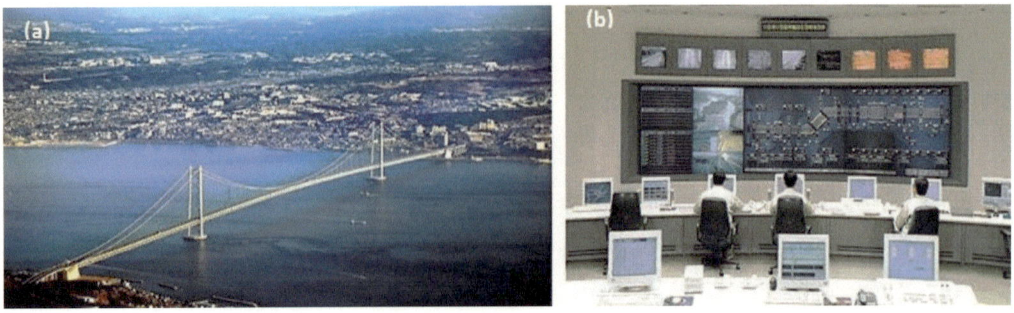

Figure 13. Akashi Kaikyo bridge: (a) look-out on the bridge; (b) 24 h a day monitoring center (source: *Kobe—Awaji—Naruto Expressway.* Honshu—Shikoku Bridge Authority 1998, advertising material).

http://dx.doi.org/10.5772/intechopen.73193

5. Case study

The *dynamic loading test* and the following *dynamic monitoring* of the Lafranconi bridge over the Danube in Bratislava (Slovakia) are shortly described in this section [38–40]. The dynamic response behavior of a prestressed concrete, seven span highway bridge (761.0 m long) was examined via DLT according to standard [24] in 1990. Excitations of bridge structure were induced by *the passage of two fully loaded, multi-axles lorries* as well as by the *rocket engines*. Applied structural measurement technique was developed for in situ testing of the bridges. The DLT results enabled to identify bridge global dynamic characteristics of the bridge, e.g., *maximum* and *RMS of displacements amplitude, natural frequencies f*(j), *mode shapes*, DAF → δ_{OBS} ($\delta_{OBS} = w_{max}/w_s$) and the *structure amplitude damping parameter* (ϑ). The obtained dynamic characteristics were compared with the numerical computed data [29] and standard prescriptions. For maximum and RMS displacements amplitude, and so on, see technical report [2].

5.1. Dynamic loading test of Lafranconi highway bridge over the Danube

The main bridge structure is composed of seven span continuous beams with one bridge frame pier (P3). The total length of the bridge was 761.0 m with spans 83.0 m + 174.0 m + 172.0 m + 4 × 83.0 m. The highway bridge consists of two independent bridges (left and right bridge) with three traffic lanes each (i.e., three in each bridge for one direction only) and sidewalks on both sides. The bridge's longitudinal section is shown in **Figure 14**. The bridge structure including multispan junctions, the test program, field measurements, and applied instrumentation are fully described in [29]. The vibration amplitudes were measured and recorded in 18 selected points. The measuring station for recording accelerometer signals (DSM-1) was situated on the top of the pier P3, **Figure 14**. The time history of vertical as well as horizontal vibration amplitudes have been registered by accelerometers ın the second and the third span of the bridge. In the other bridge spans were applied inductive displacement transducer with working

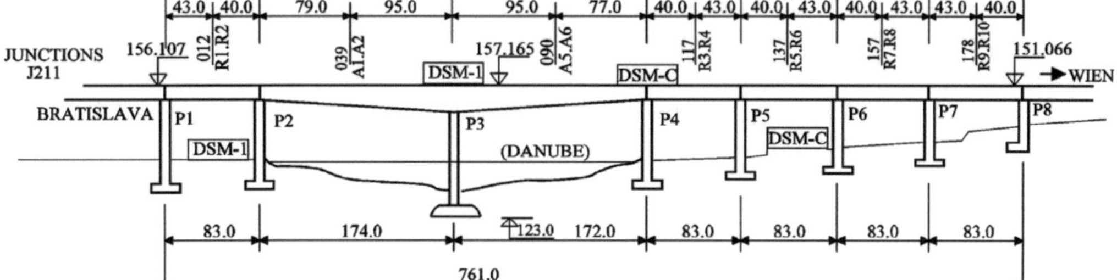

Figure 14. Longitudinal section of the Lafranconi highway bridge over the Danube in Bratislava.

range $\pm$ 40 mm. **Figure 14** shows the position of the accelerometers marked as A1, A2, A5, and A6 and transducers marked as R1–R10.

Output signals from the accelerometers were preamplified and recorded on two PC and four-channel portable tape FM recorders (BK-7005) and the *signals from the inductive displacement transducers* were recorded simultaneously by 12-channel portable tape recorder and FM recorder (BK-7005). The *DAF* have been determined by analyses of bridge amplitude vibration records from computer or tape recorders via relevant PC software pocket (*Disys*, 1990). The *frequency response spectra* (power spectrum $D_{xx}(f)$, power spectral densities $G_{xx}(f)$, cross power spectral densities $G_{xy}(f)$, etc.) have been obtained by PC spectral analysis programs and by coupled two-channel analyzer BK-2032 in the frequency range 0–10 Hz. Output signal in the form of *power spectrums* were recorded by digital recorder (BK-7400) and plotted by x–y plotter BK-2308. The bridge vibrations *ambient-ability* have been investigated by means of the correlation and spectral analysis by cross-correlation functions $R_{xy}(t)$ and coherence function $\gamma^2{}_{xy}(f)$. Examples of amplitude (a), and spectral (b) analyses results, DAF dependence on lorry velocities (c) and calculated and measured natural frequencies comparison (d) are depictured in **Figure 15**.

5.2. Bridge dynamic parameters monitoring

In this section, bridge monitoring process and results are shortly described. *Lafranconi bridge* over the *Danube* has been investigated by 24 h of bridge monitoring tests in the summer and

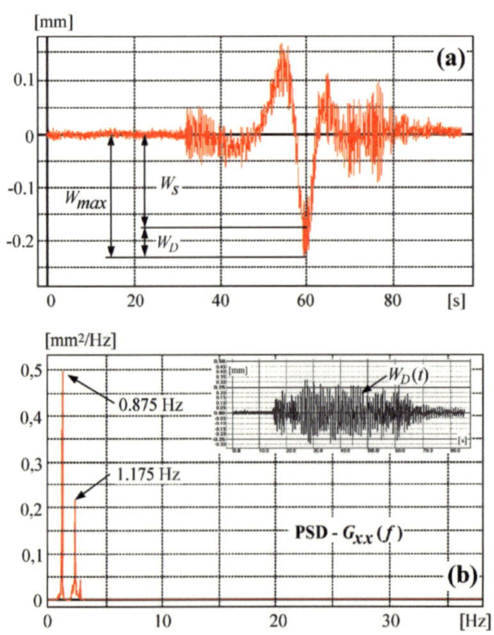

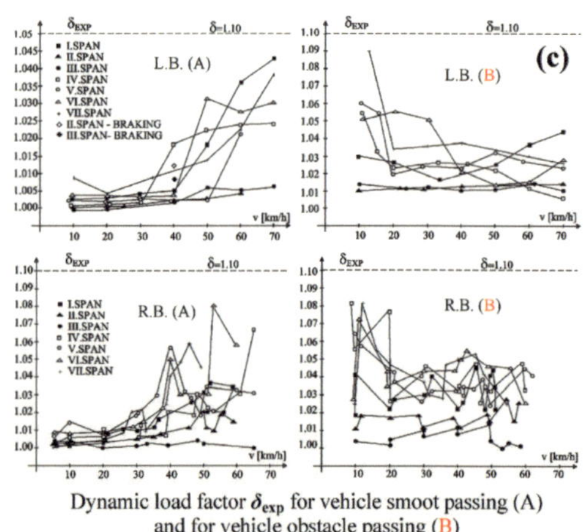

(d)

f(j)	Natural frequencies (Hz) Calculated	Measured	Exitation type
1	0.865	0.875	RE, TATRA 815 - BR
2	1.157	1.175	TATRA 815 - OBST
3	1.919	1.9	RE
4	2.042	2.148	TATRA 815
5	2.3	2.275	RE, TATRA 815 - OBST
6	2.686	2.775	RE, TATRA 815 - BR

Figure 15. Examples of amplitude (a), and spectral (b) analyses results; DAF dependence on lorry velocities (c) and calculated and measured natural frequencies comparison (d).

http://dx.doi.org/10.5772/intechopen.73193

the winter time during the years 1991–2001. The theoretical and experimental predictions of the bridge behavior and former DLT results are reported in [29, 38].

Bridge testing and experimental procedures: the bridge vibration amplitudes were measured and recorded in selected points of the second (174.0 m) and the third span (172.0 m) of the bridge. The time history of bending vertical amplitude vibration has been recorded by accelerometers at points marked as Al, A2, A5, and A6 (see also **Figure 14**) that were situated in the same position as during DLT [29]. Output signals from accelerometers were preamplified and recorded by portable computer (PC) with relevant software and hardware facilities for 24 h continuing test. The analysis of the experimental measured data has been carried out in the laboratory conditions. The records obtained in the bridge monitoring tests were investigated by using frequency analyzer BK-2034 and mentioned PC facilities. **Figure 16** shows power spectral density example of the monitoring test performed in August, 1994. The damping parameters were found by means of the 3 dB *bandwidth method* and *curve fitting techniques*. The amplitude analysis has been used to obtain *RMS* amplitude value of the bridge vibrations during the monitoring tests. As an example, **Figure 17** shows results in the form of *dominant frequency, damping* for lowest natural frequency in bending vibrations and *RMS* amplitude value from the monitoring period of years 1990–1997. A 2.7% change in frequency was observed during an year (summer-winter) but it is systematic from 1 year to the next year and is maybe due to changes in ambient temperature. The *frequencies measured at the same of the annular monitoring period have changes from year to year small and non–systematic* (coefficient of variation of about 0.01).

In comparison with the determined changes in structures natural frequency of about 30% corresponding to advanced failure observed in [31], it may be considered negligible. There are not systematic changes of structure damping but scattering of results are big [29]. These changes are maybe caused by changes in temperatures during the day; also, there are influences of changes in length of bridge which can modify support conditions and structural damping.

From monitoring results follow difference values of the *displacement RMS amplitude* measured in May 1991 in comparison with other measurements results. It was caused by both side motor traffic flows only on the left bridge. All the following monitoring measurements were performed in conditions of the one-side traffic flow on each of the both Lafranconi bridge. The changes of the *amplitude RMS value* are caused mainly by changes of the intensity of the regular motor traffic on the bridge.

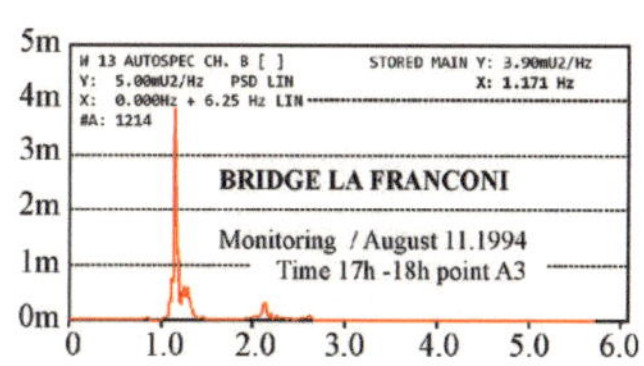

Figure 16. Power spectral density $G_{33}(f)$ of the bridge vibration displacement amplitude at point A3.

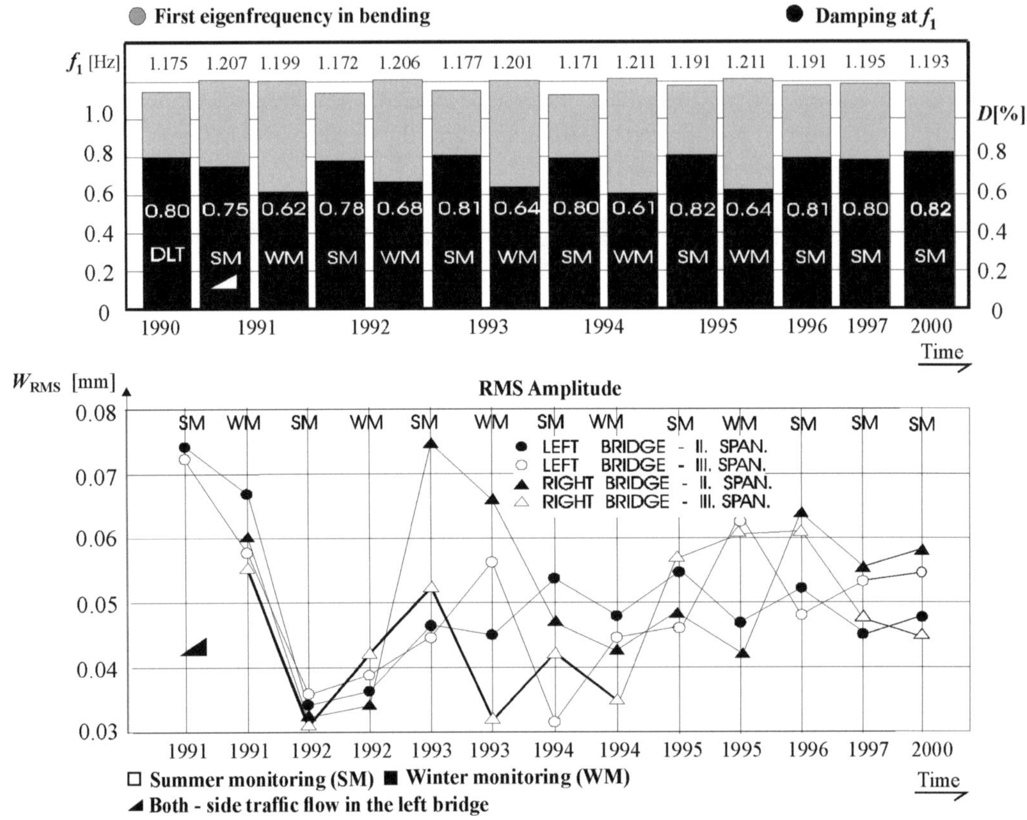

Figure 17. Changes in relative frequency, damping, and displacement RMS amplitude values during 1990–1997.

6. Conclusions

There has been a considerable amount of research conducted in the fields of bridge dynamics. From the analytical and experimental findings, the following conclusions arise:

1. Many experimental results have proved that the DAF is related to the bridges fundamental frequencies in the range 1–6 Hz. Most of the commercial lorries moving on bridges have fundamental frequencies in the range of 2–5 Hz, corresponding to the fundamental frequencies of roads and highway bridges. During the common performance of the bridges it results resonances effects.

2. Analytical and numerical models cannot reliably calculate the DAF for bridges with many specific boundaries, and initial and mechanical input parameters because of the difficulty to model them without experimental proving tests.

3. Many different formulas were suggested to evaluate the DAF from the experimental data obtained under testing vehicles (*DLT*) or regular traffic loading (*normal traffic flow*). More

researchers have used the ratio of maximum dynamic response over the maximum filtered response (e.g., deflections) as a definition of the DAF.

4. Inappropriate position of the pickups on a bridge cross section can give an unreliable experimental value of the DAF from bridge dynamic tests.

5. Full-scale testing under moving vehicle (*DLT, traffic flow*) loading is still the only economical and practical way to evaluate the *DAF* with reasonable certainty [27]. It is also suitable a reliable method for determining bridge structural dynamic properties and fully acceptable mainly for inspection purposes (even in cases of highway bridge dynamic investigation where DLF for highway bridges is not defined, e.g., in *Eurocodes*).

Acknowledgements

We kindly acknowledge the project APVV-14-0508 supported by Slovak Research and Development Agency and Slovak state budget.

Author details

Ján Benčat[1]* and Robert Kohár[2]

*Address all correspondence to: jan.bencat@gmail.com

1 Institute of Competitiveness and Innovations, University of Žilina, Žilina, Slovakia

2 Faculty of Mechanical Engineering, University of Žilina, Žilina, Slovakia

References

[1] Willis R. Experiments for determining the effects produced by causing weights to travel over bars with different velocities. In: Grey G et al., editors. Report of the Commissioners Appointed to Inquire into the Application of Iron to Railway Structures. London: W. Clowes; 1849

[2] Stokes GG. Discussion of a differential equation relating to the breaking of railway bridges. Transaction Cambridge Philosophical Society. 1849;**8**(Part 5):707-735

[3] Zimmerman H. Die Schwingungen eines Tragers mit bewegter Last. Centralblatt der Bauver waltung, Berlin. 1896;**16**(23):249-251, (23A):257-260, (24):264-266, (26):288

[4] Krylov AN. Mathematical collection of papers of the. Academy of Sciences. 1905;**61**:211

[5] Timoshenko SP. On the forced vibration of bridges. Philosophical Magazine. 1922;**6**(43): 1018, 707-735

[6] Inglis CE. A Mathematical Treatise on Vibration in Railway Bridges. Cambridge: Cambridge University Press; 1934

[7] Rao GV. Linear dynamics of an elastic beam under moving loads. Journal of Vibration and Acoustics, Transactions of the ASME. 2000;**122**(3):281-289

[8] ASCE Reports of Committee 105. ACI Journal. 1931;**27**:761

[9] Biggs JM, Suer HS, Louw JM. Vibration of simple–span–highway bridges. Transactions of ASCE. 1959;**124**(2979):291-318

[10] Looney CTG. High-speed computer applied to bridge impact. Journal of Structural. Division Proceedings ASCE. 1958;**84**(ST 5):1-41

[11] Huang T, Valetsos AS. A Study of Dynamics Response of Cantilever Highway Bridges, 206, Civil Engineering Studies, Structural Research Series. University of Illinois Press; 1960

[12] Tung TP, Goodman LE, Chen TY, Newmark NM. Highway bridge impact problems. Highway Research Board Bulletin. 1956;**124**:111-134

[13] Chaallal O, Shahawy M. Experimental evaluation of the dynamic amplification factor for evaluation of bridge performance. Research report No. ETS.DRSR.98.1. University of Quebec; 1998. p. 24

[14] Koloušek V. Civil Engineering Structures Subjected to Dynamic Load, Part I. STNL: Prague; 1967 (in Slovak)

[15] Frýba L. Vibration of Solids and Structures under Moving Loads. Prague: Academia; 1972

[16] Baťa M. Dynamics of road bridges. In: Proceedings of the First International Conference Traffic Effects on Structures and Environment (TESE 01); 1–3 December 1987; The High Tatras, Czechoslovakia; UTC Žilina. 1987

[17] Benčat J. Report on dynamic loading test results of the rail–highway bridge in Bratislava. Research Report HZ 10/a–81, UTC Žilina, Czechoslovakia; 1983. p. 226 (in Slovak)

[18] Ting EC, Genin J, Ginsberg JH. A general algorithm for moving mass problems. Journal ofSound and Vibration. 1974;**33**(l):49-59

[19] Cantieni R. Dynamic load testing of highway bridges. In: IABSE Proceedings P–75/ 84.1984

[20] He L, Reynders E, Hsu T, De Roeck G. Analysis of dynamic coupling between spans of two multi–span bridges using ambient vibration measurements. In: Proceedings of the 8th International Conference on Structural Dynamics, EURODYN 2011; 4–6 July 2011; Leuven, Belgium. pp. 1552-1558

[21] Harik I, Zhao T, Hu J. Seismic evaluation of long span bridges in Kentucky. In: Proceedings of the 8th International Conference on Structural Dynamics, EURODYN 2011; 4–6 July 2011; Leuven, Belgium. pp. 1393-1398

[22] Benčat J. Experimental analysis of dynamic characteristic of the rail–highway bridge over the Danube. In: Proceedings of the International Conference on Steel and Aluminium Structures, 8–10 July 1987; Cardiff, UK. London, UK: Elsevier Applied Science; 1987

[23] Rodrigues J, Ledesma M. Dynamic tests of a railway viaduct. In: Proceedings of the 8th International Conference on Structural Dynamics, EURODYN 2011; 4–6 July 2011; Leuven, Belgium. pp. 1393-1399

[24] Beben D. Dynamic amplification factors of corrugated Steel plate culverts. Engineering Structures. 2013;**46**:193-204. DOI: 10.1016/j. engstruct.2012.07.034

[25] Wen RK. Dynamic response of beams traversed by two-axle loads. Journal of Structural. Division; Proceedings ASCE. 1960;**86**(EM 5):2624, 91-112

[26] Benčat J, Papán D. Bridge structures dynamic analysis and vibration control. In: The 23rd International Congress on Sound and Vibration; 10–14 July, 2014; Athens, Greece. https://www.iiav.org/archives_icsv_last/2016_icsv23/content/papers/papers/full_paper_1016_20160625083552172.pdf

[27] Salawu OS, Wiliams C. Review of full-scale dynamic testing of bridge structures. Engineering Structures. 1995;**17**(2):113-112

[28] Slovak Standard No. 736209—Loading tests of bridges (in Slovak)

[29] Benčat J. Highway bridges dynamic parameters monitoring. In: Proceedings of International Symposium on Heavy Vehicle Weights and Dimensions. 1998. pp. 132-140. http://road–transport–technology.org/Proceedings/5/20/20ISHVWD/Part/203/HIGHWAY/20BRIDGES/20DYNAMIC/20PARAMETERS/20MONITORING/20–/20Bencat/20.pdf

[30] Proulx J, Paultre P, Challal O. Bridge dynamics and dynamic amplification factors. Canadian Journal of Civil Engineering. 1992;**19**:260-278

[31] Benčat J. Report of dynamic loading test of the Lafranconi highway bridge D-201-HMO over the Danube in Bratislava. Research Report HZ-1-102-90/2. UTC Žilina. 1990. p. 248

[32] Ozakan E, Main J, Jones P. Long-term measurement on a cable-stayed bridge. In: Proceedings of 21th IMAC Conference; Orlando, Florida; 2003

[33] Askegaard O, Lanso HE. Correlation between changes in dynamic properties and remaining carrying capacity. Materiaux et Constructions. 1986;**19**(109):11-20

[34] Andorsen JG. Long term observation of RC bridge [thesis]. Department of Structural. Engineering, Technical University of Denmark; 1985

[35] Casas JR, Aparicio AC. Structural damage identification from dynamic-test data. Journal of Structural Engineering. 1994;**120**(8):2437-2450

[36] Moses F, Ghosn M, Gombieski J. Report FHWA/OH-85/012, Weight-in-Motion Applied to Bridge Evaluation. USA: Case Western Reserve University Cleveland; 1985

[37] Laman JA, Nowak AS. Research Report, Fatigue Load Spectra for Steel Girder Bridges. UMCE 92–34; University of Michigan, USA. 1992

[38] Benčat J. Technical Report, Monitoring Test Results of the D-201-HMO Highway Bridge-Lafranconi. Hz-SvF-91, UTC Žilina. Slovakia. 1993 (in Slovak)

[39] Benčat J. In: Narayanan R, Roberts TM, editors. Bridges subjected to dynamic load, Structures subjected to dynamic loading. London and New York: Elsevier Applied Science; 1991

[40] Benčat J. Dynamic tests on a highway steel arch viaduct. In: Topping BHV, Iványi P, editors. Proceedings of the Twelfth International Conference on Computational Structures Technology. Stirlingshire, Scotland: Civil-Comp Press; 2014